Cosmic War Scrolls

Angels, Demons, and Interplanetary Battle in Qumran and Early Christian Writings

Unveiling Clashes of Heavenly Hosts That May Point to Alien Conflicts or Galactic Warfare

A Modern Translation

Adapted for the Contemporary Reader

Various Ancient Writers

Translated by Tim Zengerink

Table Of Contents

Preface - Message to the Reader

What If You Could Help Rebuild the Greatest Library in Human History?

Thousands of years ago, the Library of Alexandria stood as the crown jewel of human achievement — a sanctuary where the collected wisdom of every known civilization was gathered, preserved, and shared freely.

And then, it was lost.

Through fire, conquest, and the slow erosion of time, humanity lost not just books — but ideas, dreams, discoveries, and stories that could have changed the world forever.

Today, the Library of Alexandria lives again — and you are invited to be a part of its restoration.

Our mission is simple yet profound:

To rebuild the greatest library the world has ever known, and to translate all timeless works into every language and dialect, so that no seeker of knowledge is ever left behind again.

By joining our movement to rebuild the modern Library of Alexandria, you become part of an unprecedented mission:

- **Unlimited Access to the Greatest Audiobooks & eBooks Ever Written:**

  Instantly explore thousands of legendary works—Plato, Shakespeare, Jane Austen, Leo Tolstoy, and countless more. All instantly available to read or listen, placing a complete literary universe at your fingertips.

- **Beautiful Paperback & Deluxe Editions at Printing Cost**

  Own any title as an elegant paperback, deluxe hardcover, or stunning collectible boxset—offered to you at true printing cost, delivered straight to your door. Build your personal Library of Alexandria, crafted for beauty, built for durability, and worthy of proud display.

- **Fresh Translations for Modern Readers—in Every Language & Dialect**

  Enjoy timeless masterpieces reimagined in clear, contemporary language—no more outdated phrases or obscure references. Alongside the original versions, we're tirelessly translating these classics into every language and dialect imaginable, ensuring accessibility and understanding across cultures and generations.

- **Join a Global Renaissance of Literature & Knowledge**

  You directly support expanding our library, publishing deluxe editions at true cost, translating works into all global languages, and bringing humanity's greatest stories to people everywhere. By joining today, you're not just preserving a legacy of masterpieces; you set in motion a powerful wave of literary accessibility.

Become a Torchbearer of Knowledge.

Join us for free now at **LibraryofAlexandria.com**

Together, we will ensure that the light of human wisdom never fades again.

With gratitude and a shared love of knowledge,
The Modern Library of Alexandria Team

Visit:

www.libraryofalexandria.com

Or scan the code below:

Introduction

Ancient Texts, Modern Questions

What if the vivid accounts of cosmic warfare found in the Dead Sea Scrolls, apocalyptic Jewish literature, and early Christian prophecy were more than mere religious allegory? What if, hidden within the symbols and poetic flourishes of ancient scripture, there lies the memory—or premonition—of a literal interstellar conflict?

This book, Cosmic War Scrolls, invites you to confront that very possibility.

For centuries, the enigmatic texts of Qumran and related writings like Jubilees, 1 Enoch, and even the Book of Revelation have been read through spiritual or theological lenses. Scholars and laypeople alike have interpreted the "war in heaven" as a metaphor for inner moral struggle, the conflict between good and evil, or the eschatological culmination of human history. But a growing number of thinkers— from ufologists and mythologists to theoretical physicists and theologians—are reconsidering these texts with a different hypothesis: that the ancient prophets may have been describing real events, couched in the language and cosmology of their time.

After all, ancient writers didn't lack intelligence. What they lacked was vocabulary for technologies, dimensions, and forces that lay beyond the scope of their comprehension. A divine chariot might have been an unidentified flying object. A blazing angel descending from heaven might be a being not from another realm, but another star system. And the "heavenly hosts" might not be metaphors at all—but instead literal combatants in an ancient, cosmic conflict spanning worlds and millennia.

This book explores these ideas in depth, not with the intention of dismissing spiritual or symbolic interpretations, but to widen the lens. Whether one sees these texts as celestial metaphors or cosmic journalism, they present a coherent worldview: the universe is not a neutral expanse of stars, but a contested domain—alive with beings, orders, laws, and wars.

We are not the first to consider this idea. Throughout history, mystics, visionaries, and prophets have hinted at extraterrestrial presences, higher intelligences, and interdimensional battles. But the recent declassification of UFO data, the rise of theoretical models involving the multiverse and simulation theory, and a renewed cultural interest in alien life have given these ancient writings a new urgency.

Could the War Scroll's detailed battle plans actually reflect observed or inherited knowledge from a non-terrestrial conflict? Could Revelation 12's war between Michael and the dragon mirror events in another dimension—or another solar system? Could the "Watchers" in 1 Enoch be extraterrestrial beings whose genetic interference with humanity sparked the great deluge?

In the pages ahead, we'll examine ancient scrolls with fresh eyes. We'll extract their military strategies, decode their celestial terminology, and analyze their characters—angels, demons, and hybrids alike—with the curiosity of both historian and explorer. This is not simply an esoteric or theological endeavor; it is a journey through ancient literature as potential cosmic record.

Our goal is not to convince, but to open the possibility. If these texts are echoes of something more than myth, we owe it to

ourselves—and to whatever truths may lie hidden among the stars—
to read them accordingly.

Mapping the Battlefield: The Texts Within

At the heart of this book lies a collection of ancient writings—each of
which plays a role in constructing the cosmic war narrative. These are
not random selections. They were chosen for their shared motifs,
overlapping imagery, and intertextual references to spiritual beings in
conflict.

The War Scroll (1QM) is the centerpiece. Discovered among the
Dead Sea Scrolls in Cave 1 at Qumran, it outlines the "War of the Sons
of Light Against the Sons of Darkness." This is not a vague or symbolic
narrative—it includes highly detailed military strategies, formations,
calendar cycles, trumpet calls, and battle roles for both human and
angelic participants. Some of these instructions bear a striking
resemblance to modern warfare protocol, while others hint at
technologies or phenomena far beyond anything in the Second Temple
period.

The Temple Scroll, another significant Qumran text, expands the
dimensions of divine law and places significant emphasis on the purity
and sanctity of warfare. Its references to heavenly involvement in
human battles hint at a larger unseen dimension to the conflict—one
that connects human ritual to cosmic balance.

Jubilees, often called the "Lesser Genesis," rewrites early Biblical
history from a more cosmic and angel-centric point of view. It
emphasizes the role of "Watchers"—angelic beings who descended to
earth and taught humanity forbidden knowledge. Their transgressions,
offspring, and eventual judgment are central to apocalyptic theology

and could be read as a tale of ancient intervention from an advanced nonhuman species.

Revelation 12 rounds out our textual foundation. The vision of a great red dragon sweeping a third of the stars from the sky and waging war in heaven has captivated Christian audiences for centuries. But in light of the other texts, it may be more than eschatological allegory—it may be a direct reference to a cosmic uprising that has echoes in Babylonian, Zoroastrian, and Hebrew lore.

Throughout this book, these texts are not treated in isolation. They are positioned side by side, cross-referenced, and interpreted with the aid of ancient cosmologies, Second Temple angelology, and even current fringe theories involving extraterrestrial intelligence. We ask provocative questions and leave the reader with multiple interpretive paths.

Are these merely religious fantasies projected into the heavens—or are they half-remembered chronicles of a time when earth was not a backwater in the cosmos, but a battlefield?

Between Faith and Frontier: Reading with Open Eyes

A book like this necessarily dances at the edge of disciplines. It is not strictly theology, nor is it pure science fiction. It draws from archaeology, mysticism, philology, comparative religion, ancient astronaut theory, and emerging theories in cosmology. The reader should expect tension—and welcome it. That tension is the sign that we're pushing into territory that resists easy classification.

We are not trying to reduce the sacred to the scientific, nor inflate science into a new religion. Rather, we are looking for the space where they might overlap. Where ancient language might conceal modern truth. Where "heaven" might mean "off-world," and "angel" might mean "interdimensional being."

This doesn't require abandoning faith. On the contrary, many believers have found their understanding of scripture deepened—not diminished—by considering these cosmic dimensions. After all, if God created the heavens and the earth, why would His narrative be confined to just one planet?

Furthermore, it's worth noting how similar mythologies crop up across cultures. From the Hindu Devas and Asuras locked in celestial wars, to the Norse Aesir and Vanir, to the Sumerian tales of gods descending in flying machines, the pattern repeats. A heavenly realm populated by superhuman beings at war—affecting human destiny— might be a universal archetype. Or it might be universal history, preserved in different languages.

We leave that decision to you.

What we offer here is a guide through the oldest known documents that explicitly describe the machinery, strategy, and actors of cosmic conflict. You will not only read the ancient scrolls, but encounter them in conversation—with each other, with scripture, with mythology, and with modern interpretations. Each chapter opens a new door. Some lead inward, to the spiritual. Others lead outward, to the stars.

Read carefully. Read curiously. Whether you believe these stories are allegory, history, or prophecy—they speak to something that modern science is only beginning to imagine: that we are not alone, and that the universe may be more alive, more storied, and more dangerous than we ever dared to believe.

Welcome to the war among the stars.

The War Scroll
(1QM)

Introduction

A Battle Beyond Time: Unveiling the Apocalyptic Vision of the War Scroll

Among the most dramatic and enigmatic writings discovered in the caves of Qumran, The War Scroll—also known by its manuscript designation 1QM—is a document of extraordinary significance. It captures a cosmic vision of warfare between the forces of light and darkness, an apocalyptic blueprint for a final showdown in which good will ultimately triumph through divine power and purity. This unique composition reveals much about the sectarian ideology, military imagination, and eschatological hope of the Qumran community—widely believed to be the Essenes—and offers modern readers a rare glimpse into how ancient Jewish sects envisioned the end of days.

The scroll, found in Cave 1 near the Dead Sea in 1947 and designated as 1QM (Cave 1, Qumran, Milḥamah—Hebrew for "war"), consists of 19 columns of Hebrew text, written in a style that combines legalistic detail with prophetic poetry. While many other Dead Sea Scrolls offer commentary on biblical texts or instructions for daily communal living, The War Scroll is distinctive in its focus on eschatological conflict and military organization. It is both a battle plan and a theological manifesto—one that imagines not just political resistance to Rome or other earthly powers, but a divinely orchestrated confrontation between righteousness and evil on a cosmic scale.

At its core, The War Scroll envisions a prolonged war between the "Sons of Light"—the righteous members of the covenantal community—and the "Sons of Darkness," a symbolic coalition of enemies that includes the Kittim (usually interpreted as the Romans), various nations of the Gentiles, and the forces of Belial (a term often used in Qumran texts for Satan or the demonic). This war is no mere metaphor. The scroll lays out detailed instructions for military formations, priestly blessings, trumpet signals, and liturgical recitations to be used in battle. Yet, despite its tactical focus, the true power of the text lies not in its military realism, but in its apocalyptic vision: the war is ultimately God's war, and victory comes through divine intervention and ritual purity rather than brute force.

The scroll presents two main phases of warfare: a 40-year campaign of alternating victories and defeats, followed by a final, decisive battle in which the Sons of Light are saved by God's miraculous intervention. This structure reflects the community's deep sense of historical suffering and their expectation of a future age in which God would overturn worldly powers and vindicate the faithful. Echoes of the scroll's theology can be found in other Jewish apocalyptic works, such as 1 Enoch, Daniel, and 4 Ezra, as well as in the New Testament book of Revelation. Yet The War Scroll is unique in its combination of martial detail, ritual precision, and cosmic scope.

Scholars have debated whether the scroll was meant to be a literal battle plan or a symbolic framework for spiritual warfare. There is evidence to support both interpretations. On one hand, the meticulous listing of army divisions, banners, weapons, and trumpet calls suggests a real-world military aspiration—perhaps reflecting the community's hopes during the Maccabean period or the early years of Roman occupation. On the other hand, the heavy emphasis on priestly rituals, angelic assistance, and the supernatural defeat of the enemy implies that the true battlefield lies beyond human control. Either way, the scroll reveals a community obsessed with order, holiness, and the vindication of divine justice.

The world that produced The War Scroll was one of political instability, religious fragmentation, and eschatological fervor. The Qumran sect viewed itself as the remnant of Israel, the true guardians of the covenant, set apart from a corrupt Temple priesthood in Jerusalem and awaiting the coming of two messiahs: one priestly and one royal. Their vision of the end times was both terrifying and hopeful—terrifying in its scale of violence and judgment, hopeful in its promise of renewal and restoration. For them, the coming war would not only destroy the wicked but also purify the faithful, inaugurating a new age of peace, holiness, and divine rule.

In the modern era, The War Scroll has fascinated historians, theologians, military theorists, and even writers of speculative fiction. It raises profound questions about the relationship between religion and violence, the role of apocalyptic thinking in shaping human behavior, and the enduring human desire to see history as part of a larger, divinely ordered narrative. As we read this scroll today, we must resist the temptation to treat it simply as an artifact of ancient fanaticism. Instead, we are invited to enter into the worldview of a people who, in the face of overwhelming external pressures, found hope not in political accommodation but in the promise of divine victory.

The scroll's vivid language, precise organization, and theological intensity make it one of the most compelling apocalyptic texts in the Dead Sea corpus. Whether we interpret its vision of war literally, symbolically, or somewhere in between, it forces us to grapple with the darker sides of religious expectation, as well as the enduring human yearning for justice, redemption, and the ultimate triumph of good over evil.

Structure, Ritual, and Strategy in the Scroll's Vision of Warfare

At first glance, the militaristic nature of The War Scroll may seem at odds with the ascetic and pacifist image sometimes attributed to the Essenes. Yet the scroll reveals a deep congruence between war and worship, between strategy and sanctity. The document is highly structured, with an intricate organization of forces, weaponry, and ritual, indicating that its authors saw war not as a chaotic eruption of violence but as a sacred act conducted under divine instruction and heavenly command.

The scroll begins by outlining the forces involved in the conflict. The Sons of Light, representing the covenant community, are to be organized into divisions, each led by tribal commanders and supported by priests and Levites who recite prayers and offer blessings during battle. The centrality of priestly presence underscores the idea that the war is fundamentally spiritual. The priests are not passive bystanders; they are the liturgical backbone of the army, sanctifying the conflict and ensuring that divine favor is maintained.

The enemy, the Sons of Darkness, is portrayed as a vast and terrifying coalition. Chief among them are the Kittim—a term that originally referred to people from Cyprus or the Aegean but came to symbolize the oppressive forces of the Roman Empire. These enemies are described as wicked, idolatrous, and under the command of Belial, the adversary spirit. The war is thus framed as a cosmic confrontation, a reenactment of the primordial struggle between God and chaos, light and darkness, order and disorder.

The scroll then proceeds to outline the sequence of battles. The initial phase is a back-and-forth war that lasts 40 years, echoing the Israelites' time in the wilderness. Each year, a different tribe takes its turn in leading the campaign, ensuring communal participation and mirroring the sacred calendar. This period of alternating victories and defeats is not seen as failure, but as a necessary test of endurance and purification. Only after this prolonged trial does the final battle occur, when God intervenes directly, annihilates the forces of darkness, and ushers in an age of peace.

One of the most fascinating aspects of the scroll is its attention to detail. It includes instructions on the number of soldiers per unit, the types of weapons to be used, the construction of banners, and even the inscriptions to be placed on shields and trumpets. Each trumpet signal has a distinct purpose—assembly, advance, retreat, or prayer—demonstrating a highly developed communication system. This precision reflects a worldview in which every aspect of life, including warfare, must be ordered according to divine patterns.

Equally important are the prayers and psalms to be recited before and during the battles. These liturgical texts call upon God for strength, protection, and victory. They affirm that the war is not for personal gain or national glory, but for the vindication of God's name and the establishment of His kingdom on earth. These rituals blur the line between temple and battlefield, suggesting that for the Qumran sect, holiness was not confined to sacred space but extended into every realm of existence—including the realm of violence.

The scroll also emphasizes the purity of the warriors. Only those who are ritually clean, morally upright, and physically unblemished may participate. This reflects the sect's obsession with purity and separation from a corrupt world. The war is not just against external enemies, but also against internal impurity and disobedience. Victory is conditional upon the faithfulness of the community; sin and corruption can bring defeat, while righteousness ensures divine support.

This fusion of military and priestly elements reflects the Qumran belief in a dual messiah: a priestly figure from the line of Aaron and a royal warrior from the line of David. Though not explicitly named in The War Scroll, this expectation undergirds the entire vision. The war is not only fought by humans but by angelic hosts under divine leadership. In the final battle, the "great hand of God" will strike down the enemy, revealing the scroll's ultimate message: that the battle belongs to the Lord.

Apocalyptic Worldview and Enduring Significance

To understand The War Scroll is to understand the apocalyptic mindset that shaped much of Second Temple Judaism and laid the groundwork for both Rabbinic Judaism and early Christianity. Apocalypticism was not merely about predicting the future; it was a comprehensive worldview that interpreted present suffering in light of divine revelation. It offered hope in the midst of despair, clarity amid confusion, and a vision of cosmic justice that transcended human limitations.

In this worldview, history is not cyclical or aimless, but linear and purposeful. It moves inexorably toward a final climax in which God will intervene, evil will be defeated, and the righteous will be vindicated. The Qumran community believed they were living in the "end of days," a period of trial that would culminate in the fulfillment of God's promises. The war described in the scroll was not a fantasy, but a theological necessity. It gave meaning to their suffering, coherence to their communal identity, and urgency to their religious practices.

Modern readers may find this vision troubling or even dangerous, especially in light of how religious apocalypticism has sometimes fueled real-world violence. Yet it is crucial to read The War Scroll in its historical context—as the product of a marginalized community seeking to preserve its integrity, resist assimilation, and hold fast to a vision of divine justice in the face of overwhelming power.

At the same time, the scroll continues to challenge and inspire. Its vision of a cosmic battle forces us to reflect on the nature of evil, the role of divine justice, and the ethical demands of spiritual life. It raises uncomfortable but necessary questions: Can holiness be militant? Is violence ever sanctified? How do we balance faith in divine intervention with responsibility for moral action?

For scholars, The War Scroll remains a treasure trove of insights into ancient Jewish theology, liturgy, and eschatology. For spiritual seekers, it offers a window into a passionate, disciplined, and visionary community that dared to imagine the transformation of the world. And for anyone concerned with the intersection of religion and power, it serves as a sobering reminder of the potency—and peril—of apocalyptic hope.

In the end, The War Scroll is not merely about war. It is about the struggle for meaning in a world of suffering, the hope for divine justice in an age of corruption, and the faith that even amid darkness, the light will prevail. To read it is to enter a world where every trumpet blast is a prayer, every battle a liturgy, and every act of courage a sign that history is moving toward its divinely ordained goal. For those willing to hear its call, the scroll offers not only a vision of war, but a summons to holiness, vigilance, and unwavering faith in the triumph of good.

The War Scroll (1QM)

Col.1

1. For the In[structor, the Rule of] the War. The first attack of the Sons of Light shall be undertaken against the forces of the Sons of Darkness, the army of Belial: the troops of Edom, Moab, the sons of Ammon, the [Amalekites],

2. Philistia, and the troops of the Kittim of Asshur. Supporting them are those who have violated the covenant. The sons of Levi, the sons of Judah, and the sons of Benjamin, those exiled to the wilderness, shall fight against them

3. with [...] against all their troops, when the exiles of the Sons of Light return from the Wilderness of the Peoples to camp in the Wilderness of Jerusalem. Then after the battle they shall go up from that place

4. a[nd tile king of; the Kittim [shall enter] into Egypt. In his time he shall go forth with great wrath to do battle against the kings of the north, and in his anger he shall set out to destroy and eliminate the strength of

5. I[srael. Then the]re shall be a time of salvation for the People of God, and a time of dominion for all the men of His forces, and eternal annihilation for all the forces of Belial. There shall be g[reat] panic [among]

6. the sons of Japheth, Assyria shall fall with no one to come to his aid, and the supremacy of the Kittim shall cease that wickedness be overcome without a remnant. There shall be no survivors of

7. [all the Sons of] Darkness.

8. Then [the Sons of Rig]hteousness shall shine to all ends of the

world continuing to shine forth until end of the appointed seasons of darkness. Then at the time appointed by God, His great excellence shall shine for all the times of

9. e[ternity;] for peace and blessing, glory and joy, and long life for all Sons of Light. On the day when the Kittim fall there shall be a battle and horrible carnage before the God of

10. Israel, for it is a day appointed by Him from ancient times as a battle of annihilation for the Sons of Darkness. On that day the congregation of the gods and the congregation of men shall engage one another, resulting in great carnage.

11. The Sons of Light and the forces of Darkness shall fight together to show the strength of God with the roar of a great multitude and the shout of gods and men; a day of disaster. It is a time

12. of distress fo[r al]l the people who are redeemed by God. In all their afflictions none exists that is like it, hastening to its completion as an eternal redemption. On the day of their battle against the Kittim,

13. they shall g[o forth for] carnage in battle. In three lots the Sons of Light shall stand firm so as to strike a blow at wickedness, and in three the army of Belial shall strengthen themselves so as to force the retreat of the forces

14. [of Light. And when the] banners of the infantry cause their hearts to melt, then the strength of God will strengthen the he[arts of the Sons of Light.] In the seventh lot : the great hand of God shall overcome

15. [Belial and al]l the angels of his dominion, and all the men of [his forces shall be destroyed forever].

The annihilation of the Sons of Darkness and service to God during the war years.

16. [,...] the holy ones shall shine forth in support of [...] the truth for the annihilation of the Sons of Darkness. Then [...]

17. [...] a great [r]oar [...] they took hold of the implement[s of war.]

18. [...]

19. [... chiefs of the tribes ... and the priests,

20. [the Levites, the chiefs of the tribes, the fathers of the congregation ... the priests and thus for the Levites and the courses of the heads of]

Col. 2

1. the congregation's clans, fifty-two. They shall rank the chiefs of the priests after the Chief Priest and his deputy; twelve chief priests to serve

2. in the regular offering before God. The chiefs of the courses, twenty-six, shall serve in their courses. After them the chiefs of the Levites serve continually, twelve in all, one

3. to a tribe. The chiefs of their courses shall serve each man in his office. The chiefs of the tribes and fathers of the congregation shall support them, taking their stand continually at the gates of the sanctuary.

4. The chiefs of their courses, from the age of fifty upwards, shall take their stand with their commissioners on their festivals, new moons and Sabbaths, and on every day of the year.

5. These shall take their stand at the burnt offerings and sacrifices, to arrange the sweet smelling incense according to the will of God, to atone for all His congregation, and to satisfY

themselves before Him continually

6. at the table of glory. All of these they shall arrange at the time of the year of remission. During the remaining thirty-three years of the war the men of renown,

7. those called of the Congregation, and all the heads of the congregation's clans shall choose for themselves men of war for all the lands of the nations. From ail tribes of Israel they shall prepare

8. capable men for themselves to go out for battle according to the summons of the war, year by year. But during the years of remission they shall not ready men to go out for battle, for it is a Sabbath

9. of rest for Israel. During the thirty-five years of service the war shall be waged. For six years the whole congregation shall wage it together,

10. and a war of divisions shall be waged during the twenty-nine remaining years. In the first year they shall fight against Mesopotamia, in the second against the sons of Lud, in the third

11. they shall fight against the rest of the sons of Aram: Uz, Hul, Togar, and Mesha, who are beyond the Euphrates. In the fourth and fifth they shall fight against the sons of Arpachshad,

12. in the sixth and seventh they shall fight against all the sons of Assyria and Persia and the easterners up to the Great Desert. In the eighth year they shall fight against the sons

13. of Elam, in the ninth year they shall fight against the sons of Ishmael and Keturah, and during the following ten years the war shall be divided against all the sons of Ham

14. according to [their] c[lans and] their [terri]tories. During the

remaining ten years the war shall be divided against all [sons of Japhe]th according to their territories.

The description of the trumpets.

15. [The Rule of the Trumpets: the trumpets] of alarm for all their service for the [...] for their commissioned men,

16. [by tens of thousands and thousands and hundreds and fifties] and tens. Upon the t[rumpets ...]

17. [...]

18. [...]

19. [...]

20. [... they shall write ... the trumpets of]

Col. 3

1. the battle formations, and the trumpets for assembling them when the gates of the war are opened so that the infantry might advance, the trumpets for the signal of the slain, the trumpets of

2. the ambush, the trumpets of pursuit when the enemy is defeated, and the trumpets of reassembly wben the battle returns. On the trumpets for the assembly of the congregation they shall write, "The called of God."

3. On the trumpets for the assembly of the chiefs they shall write, "The princes of God." On the trumpets of the formaons they shall write, "The rule of God." On the trumpets of the men of renown [they shall write],

4. "The heads of the congregation's clans." Then when they are assembled at the house of meeting, they shall write, "The testimonies of God for a holy congregation." On the trumpets

of the camps

5. they shall write, "The peace of God in the camps of His saints." On the trumpets for their campaigns they shall write, "The mighty deeds of God to scatter the enemy and to put all those who hate

6. justice to flight and a withdrawal of mercy from all who hate God." On the trumpets of the battle formations they shall write, "Formations of the divisions of God to avenge His anger on all Sons of Darkness."

7. On the trumpets for assembling the infantry when the gates of war open that they might go out against the battle line of the enemy, they shall write, "A remembrance of requital at the appointed time

8. of God." On the trumpets of the slain they shall write, "The hand of the might of God in battle so as to bring down all the slain because of unfaithfulness." On the trumpets of ambush they shall write,

9. "Mysteries of God to wipe out wickedness." On the trumpets of pursuit they shall write, "God has struck all Sons of Darkness, He shall not abate His anger until they are annihilated."

10. When they return from battle to enter the formation, they shall write on the trumpets of retreat, "God has gathered." On the trumpets for the way of return

11. from battle with the enemy to enter the congregation in Jerusalem, they shall write, "Rejoicings of God in a peaceful return."

12. The description of the banners.

13. Rule of the banners of the whole congregation according to their formations. On the grand banner which is at the head of

all the people they shall write, "People of God," the names "Israel"

14. and "Aaron," and the names of the twelve tribes of Israel according to their order of birth. On the banners of the heads of the "camps" of three tribes

15. they shall write, "the Spirit [of God," and the names of three tribes. O]n the banner of each tribe they shall write, "Standard of God," and the name of the leader of the t[ribe]

16. of its clans. [... and] the name of the leader of the ten thousand and the names of the chief[s of ...]

17. [...] his hundreds. On the banner [...]

18. [...]

19. [...]

20. [...]

Col. 4

1. On the banner of Merari they shall write, "The Offering of God," and the name of the leader of Merari and the names of the chiefs of his thousands. On the banner of the tho[us]and they shall write, "The Anger of God is loosed against

2. Belial and all the men of his forces without remnant," and the name of the chief of the thousand and the names of the chiefs of his hundreds. And on the banner of the hundred they shall write, "Hundred

3. of God, the power of war against a sinful flesh," arid the name of the chief of the hundred and the names of the chiefs of his tens. And on the banner of the fifty they shall write, "Ended

4. is the stand of the wicked [by] the might of God," and the name

of the chief of the fifty and the names of the chiefs of his tens. And on the banner of the ten they shall write, "Songs of joy

5. for God on the ten-stringed harp," and the name of the chief of the ten and the names of the nine men in his command.

6. When they go to battle they shall write on their banners, "The truth of God," "The righteousness of God," "The glory of God," "The justice of God," and after these the list of their names in full.

7. When they draw near for battle they shall write on their banners, "The right hand of God," "The appointed time of God," "The tumult of God," "The slain of God"; after these their names in full.

8. When they return from battle they shall write on their banners, "The exaltation of God," "The greatness of God," "The praise of God," "The glory of God," with their names in full.

9. The Rule of the banners of the congregation: When they set out to battle they shall write on the first banner, "The congregation of God," on the second banner, "The camps of God," on the third,

10. "The tribes of God," on the fourth, "The clans of God," on the fifth, "The divisions of God," on the sixth, "The congregation of God," on the seventh, "Those called by

11. God," and on the eighth, "The army of God." They shall write their names in full with all their order. When they draw near for battle they shall write on their banners,

12. "The battle of God," "The recompense of God," "The cause of God," "The reprisal of God," "The power of God," "The retribution of God," "The might of God," "The annihilation by God of all the vainglorious nations." And

13. their names in full they shall write upon them. When they return from battle they shall write on their banners, "The deliverance of God," "The victory of God," "The help of God," "The support of God,"

14. "The joy of God," "The thanksgivings of God," "The praise of God," and "The peace of God."

15. [The Length of the Bann]ers. The banner of the whole congregation shall be fourteen cubits long; the banner of th[ree tribes' thir]teen cubits [long;]

16. [the banner of a tribe,] twelve cubits; the banner of ten thousand, eleve[n cubits; the banner of a thousand, ten cubits; the banner of a hu]ndred, [n]ine cubits;

17. [the banner of a fifty, ei]ght cubits; the banner of a ten, sev[en cubits . . .].

18. [...]

19. [...]

20. [...]

The description of the shields.

Col. 5

1. and on the sh[ie]ld of the Leader of the whole nation they shall write his name, the names "Israel," "Levi," and "Aaron," and the names of the twelve tribes of Israel according to their order of birth,

2. and the names of the twelve chiefs of their tribes.

The description of the arming and deployment of the divisions.

3. The rule for arranging the clivisions for war when their army is complete to make a forward battle line: the battle line shall be

24

formed of one thousand men. There shall be seven forward rows

4. to each battle line, arranged in order; the stahon of each man behind his fellow. All of them shall bear shields of bronze, polished like

5. a face mirror. The shield shall be bound with a border of plaited work and a design of loops, the work of a skillful workman; gold, silver, and bronze bound together

6. and jewels; a multicolored brocade. It is the work of a skillful workman, artistically done. The length of the shield shall be two and a half cubits, and its breadth a cubit and a half. In their hands they hold a lance

7. and a sword. The length of the lance shall be seven cubits, of which the socket and the blade constitute half a cubit. On the socket there she be three bands engraved as a border of plaited

8. work; of gold, silver, and Copper bound together like an artistically designed work. And in the loops of the de sign, on both sides of the band

9. all around, shall be precious stones, a multicolored brocade, the work of a skillful workman, artistically done, and an ear of grain. The socket shall be grooved between the bands like

10. a column, artistically done. The blade shall be of shining white iron, the work of a skillful workman, artistically done, and an ear of grain of pure gold inlaid in the blade; tapered towards

11. the point. The swords shall be of refined iron, purified in the furnace and polished like a face mirror, the work of a skillful workman, artistically done, with figures of ears of grain

12. of pure gold embossed on both sides. The borders shall go straight to the point, two on each side. The length of the sword

shall be a cubit

13. and a half and its width four fingers. The scabbard shall be four thumbs wide and four handbreadths up to the scabbard. The scabbard shall be tied on either

14. side with thongs of five handbreadths. The handle of the sword shall be of choice horn, the work of a skillful workman, a varicolored design with gold and silver and precious stones.

15.

16. And when the [... take their] stand, they shall arrange seven battle lines, one behind the other

17. [...] and there shall be a space [between ... t]hirty cubits, where the infan[try] shall stand

18. [...] forward [...]

19. [...]

20. [. . . they shall sling]

Col. 6

1. seven times, and return to their position. After them, three divisions of infantry shall advance and stand between the battle lines. The first division shall heave into

2. the enemy battle line seven battle darts. On the blade of the first dart they shall write, "Flash of a spear for the strength of God." On the second weapon they shall write,

3. "Missiles of blood to fell the slain by the wrath of God." On the third dart they shall write, "The blade of a sword devours the slain of wickedness by the judgment of God."

4. Each of these they shall throw seven times and then return to their position. After these, two divisions of infantry shall march

forth and stand between the two battle lines,

5. the first division equipped with a spear and a shield and the second division with a shield and a sword; to bring down the slain by the judgment of God, to subdue the battle line

6. of the enemy by the power of God, and to render recompense for their evil for all the vainglorious nations. So the Kingship shall belong to the God of Israel, and by the holy ones of His people He shall act powerfully.

The description of the deployment of the cavalry.

7. Seven rows of horsemen shall also take position at the right and at the 1eft of the battle line. Their ranks shall be positioned on both sides, seven hundred

8. horsemen on one side and seven hundred on the other. Two hundred horsemen shall go out with one thousand men of the battle line of the infantry, and thus

9. they shall take position on all sides of the camp. The total being four thousand six hundred men, and one thousand four hundred cavalry for the entire army arranged for the battle line;

10. fifty for each battle line. The horsemen with the cavalry of the men of the entire army, will be six thousand; five hundred to a tribe. All the cavalry that go out

11. to battle with the infantry shall ride stallions; swift, responsive, unrelenting, mature, trained for battle,

12. and accustomed to hearing noises and seeing all kinds of scenes. Those who ride them shall be men capable in battle, trained in horsemanship, the range

13. of their age from thirty to forty-five years. The horsemen of the army shall be from forty to fifty years old, and they

14. [...], helmets and greaves, carrying in their hands round shields and a lance eig[ht cubits long, ...]

15. [...] and a bow and arrows and battle darts, all of them prepared in [...]

16. [...] and to shed the blood of their guilty slain. These are the [...]

17. [...]

18. [...]

19. [...]

The recruitment and age of the soldiers.

Col. 7

1. and the men of the army shall be from forty to fifty years old. The commissioners of the camps shall be from fifty to sixty years old. The officers

2. shall-also be from forty to fifty years old. All those who strip the slain, plunder the spoil, cleanse the land, guard the arms,

3. and he who prepares the provisions, all these shall be from twenty-five to thirty years old. No youth nor woman shall enter their encampments from the time they leave

4. Jerusalem to go to battle until their return. No one crippled, blind, or lame, nor a man who has a permanent blemish on his skin, or a man affected with ritual uncleanness of

5. shis flesh; none of these shall go with them to battle. All of them shall be volunteers for battle, pure of spirit and flesh, and prepared for the day of vengeance. Any

6. man who is not ritually clean in respect to his genitals on the day of battle shall not go down with them into battle, for holy angels are present with their army. There shall be a distance

7. between all their camps and the latrine of about two thousand cubits, and no shameful nakedness shall be seen in the environs of all their camps.

8.

The ministry of the priests and Levites.

9. When the battle iines are arrayed against the enemy, battle line against battle line, there shall go forth from the middle opening into the gap between the battle lines seven

10. priests of the sons of Aaron, dressed in fine wlute linen garments: a linen tunic and linen breeches, and girded with a linen sash of twined fine linen, violet,

11. purple, and crimson, and a varicolored design, the work of a skillful workman, and decorated caps on their heads; the garments for battle, and they shall not take them into the sanctuary.

12. The one priest shall walk before all the men of the battle line to encourage them for battle. In the hands of the remaining six shall be

13. the trumpets of assembly, the trumpets of memorial, the trumpets of the alarm, the trumpets of pursuit, and the trumpets of reassembly. When the priests go out

14. into the gap between the battle lines, seven Levites shall go out with them. In their hands shall be seven trumpets of rams' horns. Three officers from among the Levites shall walk before

15. the priests and the Levites. The priests shall blow the two trumpets of assem[bly ... of ba]ttle upon fifty shields,

16. and fifty infantrymen shall go out from the one gate and [...] Levites, officers. With

17. each battle line they shall go out according to all [this] o[rder....
    men of the] infantry from the gates

18. [and they shall take positi]on between the two battle lines, and
    [...] the bat[tle]

19. [...]

20. [...]

Col. 8

1. the trumpets shall blow continually to direct the slingmen until
   they have completed hurling seven

2. times. Afterwards the priests shall blow on the trumpets of
   return, and they shall go along the side of the first battle line

3. to take their position. The priests shall blow on the trumpets of
   assembly, and

4. the three divisions of infantry shall go out from the gates and
   stand between the battle lines, and beside them the cavalrymen,

5. at the right and at the left. The priests shall blow on their
   trumpets a level note, signals for the order of battle.

6. And the columns shall be deployed into their formations, each
   to his position. When they have positioned themselves in three
   formations,

7. the priests shall blow for them a second signal, a low legato
   note, signals for advance, until they draw near to

8. the battle line of the enemy and take hold of their weapons.
   Then the priests shall blow on the six trumpets

9. of the slain a sharp staccato note to direct the battle, and the
   Levites and all the people with rams' horns shall blow

10. a great battle alarm together in order to melt the heart of the enemy. With the sound of the alarm,

11. the battle darts shall fly out to bring down the slain. Then the sound of the rams' horns shall quiet, but on the tru[m]pets

12. the priests shall continue to blow a sharp staccato note to direct the signals of battle until they have hurled into the battle line

13. of the enemy seven times. Afterwards, the priests shall blow for them the trumpets of retreat,

14. a low note, level and legato. According to this rule the [pr]iests shall blow for the three divisions. When

15. the first division throws, the [priests and the Levites and all the people with rams'] horns shall blow a great alarm

16. to direct the battle until they have hurled seven times. Afterwards,] the priests [shall blow] for them

17. on the trumpets of retreat ... and they shall take their stan]d in their positions in the battle line,

18. [...] and shall take up position

19. [... the sl]ain,

20. [and all the people with rams' horns shall blow a very loud battle alarm, and as the sound goes out]

Col. 9

1. their hands shall begin to bring down the slain, and all the people shall quiet the sound of alarm, but the priests shall continue sounding on the trumpets

2. of the slain to direct the fighting, until the enemy is defeated and turns in retreat. The priests shall blow the alarm to direct the battle,

3. and when they have been defeated before them, the priests shall blow the trumpets of assembly, and all the infantry shall go out to them from the inidst of

4. the front battle lines and stand, six divisions in addition to the division which is engaged in battle: altogether, seven battle lines, twenty-eight thousand

5. soldiers, and six thousand horsemen. All these shall pursue in order to destroy the enemy in God's battle; a total annihilation

6. The priests shall blow for them the trumpets of pursuit, and they shall divide themselves for a pursuit of annihilation against all the enemy. The cavalry

7. shall push the enemy back at the flanks of the battle until they are destroyed. When the slain have fallen, the priests shall continue blowing from afar and shall not enter

8. into the midst of the slain so as to be defiled by their unclean blood, for they are holy. They shall not allow the oil of their priestly anointment to be profaned with the blood

9. of the vainglorious nations.

The description of the maneuvers of the battle divisions.

10. Rule for changing the order of the battle divisions, in order to arrange their position against [...] a pincer movement and towers,

11. lien arc and towers, and as it draws slowly forward, then the columns and the flanks go out from the [t]wo sides of the battle line [that]

12. the enemy might become discouraged. The shields of the soldiers of the towers shall be three cubits long, and their lances eight cubits l[on]g. The towers

13. shall go out from the battle line with one hundred shields on a side. F[or] they shall surround the tower on the three frontal sides,

14. three hundred shields in all. There shall be three gates to a tower, one on [the right and] one on the left. Upon all the shields of the tower soldiers

15. they shall write: on the first, "Mi[chae]l," [on the second, "Gabriel," on the third,] "Sariel," and on the fourth "Raphael."

16. "Michael" and "Gabriel" on [the right, and "Raphael" and "Raphael" on the left.

17. And [...] for to the four [... They] shall establish an ambush for the [battle line] of [...]

18. and [... they shall fal]l on the s[lain ...]

19. [...]

20. [...]

The address of the chiefpriest.

Col. 10

1. of our camps, and to keep ourselves from any shameful nakedness, and he (Moses) told us that You are in our midst, a great and awesome God, plundering all of

2. our enemies befo[re u]s. He taught us from of old through our generations, saying, when you approach the battle, the priest shall stand and speak unto the people,

3. saying, "Hear O Israel, you are approaching the battle against your enemies today. Do not be afraid nor fainthearted.

4. Do not trem[ble, no]r be terrified because of them, for your God goes with you, to fight for you against your enemies, and

to save

5. you" (Deut. 20:2-4) Our [officers shall speak to all those prepared for battle, those Willing of heart, to strengthen them by the might of God, to turn back all

6. who have who have lost heart, and to strengthen all the valiant warriors together. They shall recount that which You slpoke] by the hand of Moses, saying: "And when there is a war

7. in your land against the adversary who attacks you, then yo[u] shall sound an alarm with the trumpets that you might be remembered before your God

8. and be saved from your enemies (Num. 10:9)

The prayer of the chiefpriest.

Who is like You, O God of Israel, in he[av]en and on earth, that he can perform in accordance with Your great works

9. and Your great strength. Who is like Your people Israel, whom You have chosen for Yourself from all the peoples of the lands;

10. the people of the saints of the covenant, learned in the statutes, enlightened in understan[ding ...] those who hear the glorious voice and see

11. the holy angels, whose ears are open; hearing deep things. [O God, You have created] the expanse of the skies, the host of luminaries,

12. the task of spirits and the dominion of holy ones, the treasures of [Your] gl[ory . . .] clouds. He who created the earth and the limits of her divisions

13. into wilderness and plain, and all her offspring, with fhe fru[its ...], the circle of the seas, the sources of the rivets, and

34

the rift of the deeps,

14. wild beasts and winged creatures, the form of man and the gener[ations of] his [see]d, the confusion of language and the separation of peoples, the abode of clans

15. and the inheritance of the lands, [... and] holy festivals, courses of years and times of

16. eternity. [...] these we know from Your understanding which [...]

17. [...] Your [ears] to our cry, for [...]

18. [...] his house [...]

19. [...]

20. [...]

Col. 11

1. Truly the battle is Yours, and by the strength of Your hand their corpses have been broken to pieces, without anyone to bury them. Indeed, Goliath the Gittite, a mighty man of valor,

2. You delivered into the hand of David, Your servant, because he trusted in Your great name and not in sword and spear. For the battle is Yours.

3. He subdued the Philistines many times by Your holy name. Also by the hand of our kings You rescued us many times

4. because of Your mercy; not according to our works, for we have acted wickedly, nor for the acts of our rebelliousness. The battle is Yours, the strength is from You,

5. it is not our own. Neither our power nor the strength of our hand have done valiantly, but rather by Your power and the strength of Your great valor. Jus[t as You told

6. us in time past, saying: "There shall come forth a star out of Jacob, a scepter shall rite out of Israel, and shall crush the forehead of Moab and tear down all sons of Sheth,

7. and he shall descend of Jacob and shall destroy the remnant from the city, and the enemy shall be a possession, and Israel shall do valiantly (Num. 24:17-19). By the hand of Your anointed ones,

8. seers of things appointed, You have told us about the ti[mes] of the wars of Your hands in order that You may glorify Yourself {fight} among our enemies, to bring down the hordes of Belial, the seven

9. vainglorious nations, at the hand of the oppressed whom You have redeemed [with powe]r and retribution; a wondrous strength. A heart that melts shall be as a door of hope. You will do to them as You did to Pharaoh

10. and the officers of his chariots in the Red Sea. You will ignite the humble of spirit like a fiery torch of fire in a sheaf, consuming the wicked. You shall not turn back until

11. the annihilation of the guilty. In time past You foretold [the app]ointed time for Your handis powerful work against the Kittim, saying: And Assyria shall fall by a sword not of man, and a sword,

12. not of men, shall consume him (Isa. 31: 8).

13. For into the hand of the oppressed You will deliver the [ene]mies of all the lands; into the hands of those who are prostrate in the dust, in order to bring down all mighty men of the peoples, to return the recompense

14. of the wicked on the head of [...], to pronounce the just judgment of Your truth on all sons of man, and to make for

Yourself an everlasting name among the people.

15. [...] the wars, and to show Yourself great and holy before the remnant of the nations, so that [they] may know [that]

16. [You are God ... when You] carry out judgments on Gog and on all his company that are assembled about us ..]

17. [...], for You will do battle against them from the heave[ns ...]

18. [...] upon them for confusion [...]

19. [...]

20. [...]

Col. 12

1. For You have a multitude of holy ones in the heavens and hosts of angels in Your exalted dwelling to praise Your name. The chosen ones of the holy people

2. You have established for Yourself in a [community. The nu]mber (or The b]ook) of the names of all their host is with You in Your holy dwelling, and the n[umber of the holy one]s is in the abode of Your glory.

3. Mercies of blessing [...] and Your covenant of peace You engraved for them with a stylus of life in order to reign o[ver them]: for all time,

4. commissioning the hos[ts of I Your [e]lect by their thousands and tens of thousands together with Your holy ones [and] Your angels, and directing them

5. in battle [so as to condemn] the earthly adversaries by trial with Your judgments. With the elect of heaven [they] shall prev[ail].

6.

7. And You, O God, are awe[some] in the glory of Your dominion, and the company of Your holy ones is in our midst for etern[al] support. We [shall direct our contempt at kings, derision

8. and disdain at mighty men. For the Lord is holy, and the King of Glory is with us together with the holy ones. Migh[ty men and] a host of angels are with our commissioned forces.

9. The Hero of Wa[r] is with our company, and the host of His spirits is with our steps Our horsemen are [as] the clouds and as the mist covering the earth,

10. and as a steady downpour shedding judgment on all her offspring. Rise up, O Hero, take Your captives, O Glorious One, take

11. Your plunder, O You who do valiantly. Lay Your hand upon the neck of Your enemies, and Your foot upon the backs of the slain. Crush the nations, Your adversaries, and may Your sword

12. devour guilty flesh. Fill Your land with glory, and Your heritance with blessing. An abundance of cattle in Your fields; silver and gold and precious

13. stones in Your palaces. O Zion, rejoice greatly, and shine with joyful songs, O Jerusalem. Rejoice, all you cities of Judah, open

14. your gate[s] forever that the wealth of the nations might be brought to you, and their kings shall serve you. All they that oppressed you shall bow down to you, and the dust

15. [of your feet they shall lick. O daughter]s of my people shout out with a voice of joy, adorn yourselves with ornaments of glory Rule over the ki[ngdom of the],

16. [... and I]srael to reign eternally.

17. [...] them the mighty men of war, O Jerusalem [...]

18. the exalt]ed above the heavens, O Lord, [and let Your glory be above all the earth ...]

19. [...]

The blessings of the war recited by all the leaders after the victory.

20. [... And then the Chief Priest shall stand]

Col. 13

1. land his brothers the [pr]iests, the Levites, and all the elders of the Army with him. They shall bless from their position, the God of Israel and all His works of truth, and they shall curse

2. [Beli]al there and all the spirits of his forces. And they shall say response: "Blessed is the God of Israel for all His holy purpose and His works of truth. And blessed are

3. those who serve Him richteously, who know Him by faith.

4. And cursed is Belial for his contentious purpose, and accursed for his reprehensible rule. And cursed are all the spirits of his lot for their wicked purpose.

5. Accursed are they for all their filthy dirty service. For they are the lot of darkness, but the lot of God is light

6. [eterna]l.

7. Y[o]u are the God of our fathers. We bless Your name forever, for we are an [eter]na[l] people. You made a covenant with our fathers, and will establish it for their seed

8. throughout the ages of eternity. In all the testimonies of Your glory there has been remembrance of Your [kindness] in our midst as an assistance to the remnant and the survivors for the sake of Your covenant

9. and to re[count] Your works of truth and the judgments of

Your wondrous strength. And You, [O God], created us for Yourself as an eternal people, and into the lot of light You cast us

10. in accordance with Your truth. You appointed the Prince of Light from of old to assist us, for in [His] l[ot are all sons of righteous]ness and all spirits of truth are in his dominion. You yourself

11. made Belial for the pit, an angel of malevolence, his [dominio]n is in darkne[ss] and his counsel is to condemn and convict. All the spirits

12. of his lot -- the angels of destruction-- walk in accord with the rule of darkness, for it is their only [des]ire. But we, in the lot of Your truth, rejoice in

13. Your mighty hand. We rejoice in Your salvation, and revel in [Your] hel[p and] Your [p]eace. Who is like You in strength, O God of Israel, and yet

14. Your mighty hand is with the oppressed. What angel or prince is like You for [Your] effe[ctual] support, [fo]r of old You appointed for Yourself a day of gre[at battle ...]

15. [...] to [sup]port truth and to destroy iniquity, to bring darkness low and to lend might to light, and to [...]

16. [...] for an eternal stand, and to annihilate all the Sons of Darkness and bring joy to [al]l [the Sons of Light ...]

17. [...]

18. [... f]or You Yourself designated us for an app[ointed time ...]

19. [...]

20. [...]

Col. 14

1. like the fire of His fury against the idols of Egypt. The blessings of the war recited by all the leaders in the morning before the battle.

2. After they have withdrawn from the slain to enter the camp, all of them shall sing the hymn of return. In the morning they shall wash their clothes, cleanse themselves

3. of the blood of the sinful bodies, and return to the place where they had stood, where they had formed the battle line before the slain of the enemy fell. There they shall all bless

4. the God of Israel and joyously exalt His name together. They shall say in response: "Blessed is the God of Israel, who guards loving-kindness for His covenant and the appointed times

5. of salvation for the people He redeems. He has called those who stumble unto wondrous [accomplishment]s, and He has gathered a congregation of nations for annihilation without remnant in order to raise up in judgment

6. he whose heart has melted, to open a mouth for the dumb to sing [God's] mighty deeds, and to teach feeble [hands] warfare. He gives those whose knees shake strength to stand,

7. and strengthens those who have been smitten from the hips to the shoulder. Among the poor in spirit [...] a hard heart, and by those whose way is perfect shall all wicked nations come to an end;

8. there will be no place for all their mighty men. But we are the remn[ant of Your people. Blessed is] Your name, O God of loving-kindness, the One who kept the covenant for our forefathers. Throughout

9. all our generations You have made Your mercies wondrous for

the rem[nant of the people] during the dominion of Belial. With all the mysteries of his hatred they have not led us astray

10. from Your covenant. His spirits of destruction You have driven [away from us. And when the me]n of his dominion [condemned themselves], You have preserved the lives of Your redeemed. You raised up

11. the fallen by Your strength, but those who are great in height You will cut dow[n to humble them. And] there is no rescuer for all their mighty men, and no place of refuge for their swift ones. To their honored men

12. You will return shame, and all [their] vain existence [shall be as not]hing. But we, Your holy people, shall praise Your name for Your works of truth.

13. Because of Your mighty deeds we shall exalt [your splendor in] epochs and appointed times of eternity, at the beginning of day, at night

14. and at dawn and dusk. For Your [glorio]us p[urpose] is great and Your wondrous mysteries are in [Your] high heavens, to [raise u]p those for Yourself from the dust

15. and to humble those of the gods.

16. Rise up, rise up, O God of gods, and raise Yourself in power, [O King of Kings ...]

17. let all the Sons of Darkness [scatter from before You.] Let the light of Your majesty shi[ne forever upon gods and men, as a fire burning in the dark places of the damned]

18. Let it burn [the damned of Sh]eol, as an [eternal] burning [among the transgressors ... in all the appointed times of eternity.]

19. [They shall repeat all the thanksgiving hymns of battle there and

then return to their camps]

20. [...]

Col. 15

1. For it is a time of distress for Isra[el, a fixed t]ime of battle against all the nations. The purpose of God is eternal redemption,

2. but annihilation for al1 nations of wickedness. All those pr[epared] for battle shall set out and camp opposite the king of the Kittim and all the forces

3. of Belial that are assembled with him for a day [of vengeance] by the sword of God.

The final battle the first engagement.

4. Then the Chief Priest shall stand, and with him his brothers the p[riests], the Levites, and all the men of the army. He shall read aloud

5. the prayer for the appointed time of battle, as is written in the boo]k Serekh Itto (The Rule of His Time), including all the words of their thanksgivings. Then he shall form there

6. all the battle lines, as writ[ten in the Book of the War. Then the priest appointed for the time of vengeance by

7. all his brothers shall walk about and encourage [them for the battl]e, and he shall say in response: "Be strong and courageous as warriors.

8. Fear not, nor be discoura[ged and let not y]our [heart be faint.] Do not panic, neither be alarmed because of them. Do not

9. turn back nor [flee from the]m. For they are a wicked congregation, all their deeds are in darkness;

10. it is [their] desire. [They have established al]l their refuge [in a lie], their strength is as smoke that vanishes, and all

11. their vast assembly [is as chaff which blows away ... de]solation, and shall not be found. Every creature of greed shall wither quickly away

12. [like a flow]er at ha[rvest time ... Come,] strengthen yourselves for the battle of God, for this day is an appointed time of battle

13. [for G]od against all the n[ations, ... judgm]ent upon all flesh. The God of Israel is raising His hand in His wondrous [streng]th

14. [against all the spirits of wick[edness ... m]ighty ones of the gods are girding themselves for battl[e, and] the formation[s of the3 h[o]ly ones ·

15. [are rea]dying themselves for a day of [vengeance ...]

16. the God of I[srae]l [...]

17. to remove Bel[ial ...]

18. in his hell [...]

19. [...]

20. [...]

Col. 16

1. until every source [of ...] is come to an end. For] the God of Israel has called out a sword against all the nations, and by the holy ones of His people He will do mightily."

2. They shall carry out all this Rule [on] that [day] at the place where they stand opposite the camps of the Kittim. Then the priests shall blow for them the trumpets

3. of remembrance. The gates of w[ar] shall open, [and] the infantry shall go out and stand in columns between the battle lines. The priests shall blow for them

4. a signal for the formation and the columns [shall deplo]y at the sound of the trumpets until each man has taken his station. Then the priests shall blow for them

5. a second signal: [signs for confron]tation. When they stand near the battle line of the Kittim, within throwing range, each man shall raise his hand with his weapon of

6. war. Then the six [priests shall blow on the tr]umpets of the slain a sharp staccato note to direct the fighting. The Levites and the all the people with

7. rams' horns shall blow [a battle signa]l, a loud noise. As the sound goes forth, the infantry shall begin to bring down the slain of the Kittim, and all

8. the people shall cease the signal, [but the priest]s shall continue blowing on the trumpets of the slain and the battle shall prevail against the Kittim.

The final battle the second engagement.

9. When [Belial] prepares himself to assist the Sons of Darkness, and the slain among the infantry begin to fall by God's mysteries and to test by these mysteries all those appointed for battle,

10. the priests shall blow the trumpets of assembly so that another battle line might go forth as a battle reserve, and they shall take up position between the battle lines.

11. For those employed in battle they shall blow a signal to return. Then the Chief Priest shall approach and stand before the battle line, and shall encourage

12. their heart by [the wondrous might of God and] fortify their hands for His battle.

13. And he shall say in response: ["Blessed is God, for] He tests the he[ar]t of His people in the crucible. And not [...] have your slain [...]. For you have obeyed from of old

14. the mysteries of God. [Now as for you, take courage and stand in the gap, do not fear when God strengthens ...]

15. [...]

16. [...]

17. [...]

18. [...]

Col. 17

1. land He shall appoint their retribution with burning [...] those tested by the crucible. He shall sharpen the implements of war, and they shall not become blunt until [all the nations of] wickedness [come to an end].

2. But, as for you, remember the judgment [of Nadab and Abi]hu, the sons of Aaron, by whose judgment God showed Himself holy before [all the people. But Eleazar]

3. and Ithamar He preserved for Himself for an eternal covenant [ofpriesthood].

4. But, as for you, take courage and do not fear them [... for] their end is emptine and their desire is for the void. Their support is without st[rength] and they do not [know that from the God] of

5. Israel is all that is and that will be. He [...] in all which exists for eternity. Today is His appointed time to subdue and to

humiliate the prince of the realm

6. of wickedness. He will send eternal support to the company of His redeemed by the power of the majestic angel of the authority of Michael. By eternal light

7. He shall joyfully light up the covenant of Israel peace and blessing for the lot of God, to exalt the authority of Michael among the gods and the dominion

8. of Israel among all flesh. Righteousness shall rejoice on high, and all sons of His truth shall rejoice in eternal knowledge. But as for you, O sons of His covenant,

9. take courage in God's crucible, until He shall wave His hand and complete His fiery trials; His mysteries concerning your existence."

The final battle the third engagement.

10. And after these words the priests shall blow for them a signal to form the divisions of the battle line. The columns shall be deployed at the sound of the trumpets,

11. until each man has taken his station. Then the priests shall blow another signal on the trumpets, signs for confrontation. When

12. the infa[ntry] has approached [the battle] line of the Kitt[im], within throwing range, each man shall raise his hand with his weapon. Then the priests shall blow on the trumpets

13. of the slain [and the Levites and the al]l the people with rams' horns shall sound a signal for battle. The infantry shall attack the army

14. of the Kittim, [and as the soun]d [of the si]gnal [goes forth], they shall begin to bring down their slain. Then all the people shall still the sound of the signal, while the priests

15. continuously blow on [the trumpets of the slain], and the bat[tl]e p[revail]s against the K[ittim, and the troops of Belia]l are defeated before them.

16. Thus in the th[ird] lot [...] to fall slain [...]

The final battle the fourth, fifth, and sixth engagements. Nothing of these engagements is preserved.

The final battle the seventh engagement.

Col. 18

1. [and in the seven]th [log, when the great hand of God shall be lifted up against Belial and against all the fo[rc]es of his dominion for an eternal slaughter

2. [...] and the shout of the holy ones when they pursue Assyria. Then the sons of Japheth shall fall, never to rise again, and the Kitum shall be crushed without

3. [remnant and survivor. So] the God of Israel shall raise His hand against the whole multitude of Belial. At that time the priests shall sound a signal

4. [on the six trumpegs of remembrance, and all the battle formations shall be gathered to them and divide against all the ca[mps of the Ki]ttim

5. to completely destroy them. [And] when the sun hastens to set on that day, the Chief Priest and the priests and the [Levites] who are

6. with him, and the chiefs [of the battle lines and the men] of the army shall bless the God of Israel there. They shall say in response: Blessed is Your name, O God [of god]s, for

7. You have done wondrous things for Your people, and have

kept Your covenant for us from of old. Many times You have opened the gates of salvation for us

8. for the sak[e of Your co]venant. [And You provided f]or our affliction in accord with Your goodness toward us. You, O God of righteousness, have acted for the sake of Your name.

9.

Thanksgiving for final victory.

10. [...] You have done w]onders upon wonders with us, but from of old there has been nothing like it, for You have known our appointed time. Today [Your] power has shined forth

11. for us, [and] You [have shown] us the hand of Your mercies with us in eternal redemption, in order to remove the dominion of the enemy, that it might be no more; the hand of Your strength.

12. In bat[tle You shall show Yourself strong aga]inst our enemies for an absolute slaughter. Now the day is pressing upon us [to] pursue their multitude, for You

13. [...] and the heart of warriors You have broken so that no one is able to stand. Yours is the might, and the battle is in Your hand, and there is no

14. [God like You ...] Your [...] and the appointed times of Your will, and reprisal [...] Your [enemie]s, and You will cut of from [...] is

15. [...]

16. [...]

17. [...]

18. [...]

19. [...]

20. [... And we shall direct our contempt at kings,]

Col. 19

1. [derision and disdain at mi]ghty men. For our Majestic One is holy. The King of Glory is with us and the h[ost of His spirits is with our steps. Our horsemen are]

2. [as the clouds and as the mis]t covering the earth; as a steady downpour shedding judgment on all her offspring. Rise up, O Hero,]

3. [Take Your captives, O Glorious One, and ta]ke Your plunder, O You Who do valiantly. Lay Your hand upon the neck of Your enemies, and Your fo[o]t [upon the backs of

4. [the slain. Crush the nations, Yo]ur [adversaries,] and let Your sword devour flesh. Fill Your land with glory, and Your inheritance with blessing. An ab[undance of cattle is] s[in Your fields,

5. silver and gold] in Your palaces. O Zion, rejoice greatly, and rejoice, all you cities of Ju[dah. Open]

6. [your gates forever, so that the wealth of the nations [might be brought to you, and their kings shall serve you. All they that oppressed] you shall bow down to you,

7. [and they shall lick the dust of your feet. O dau]ghters of my [peo]ple, burst out with a voice of joy. Adorn yourselves with ornaments of glory, and r[ule] over the ki[ngdom of the ...]

8. [...] Your [...] and Israel for an [egernal dominion.

Ceremony after the final battle.

9. [Then they shall gather] in the camp that n[ig]ht for rest until

the morning. In the morning they shall come to the p[la]ce of the battle line,

10. [where the mi]ghty men of the Kittim [fell], as well as the multitude of Assyria, and the forces of all the nations that were assembled unto them, to see whether [the mu]ltitude of slain [are dead]

11. [with none to bury them; those who] fell there by the sword of God. And the Hi[gh] Priest shall approach there [with] his [depu]ty, his brothers [the priests,]

12. [and the Levites with the Leader] of the battle, and all the chiefs of the battle lines and [their officers ...]

13. [... together. When they stand before the s]lain of the Kitt[im, they shall pr]aise there the God [of Israel. And they shall say in response: ...]

14. [... to God most high and ...]

The Book of Jubilees

Introduction

The Book of Jubilees, also known as "The Little Genesis," is an ancient Jewish text that expands on the stories in Genesis and Exodus. It was likely written between the 2nd and 1st centuries BCE and retells biblical history using a unique system of time—dividing events into periods of forty-nine years, called jubilees. This structure provides a detailed timeline from creation up to the moment when God gave the law at Mount Sinai. The book also gives insight into ancient Jewish beliefs, laws, and traditions.

What makes Jubilees unique is its strong focus on the idea that God's laws came directly from Him, the importance of the Sabbath, and the role of angels in delivering God's messages. The text reflects the beliefs and customs of a particular Jewish group, possibly connected to the Essenes. While it is not included in the Hebrew Bible, it is considered a sacred text in the Ethiopian Orthodox Church and is valued for its deep religious and historical meaning.

This important book acts as a link between the Bible and later writings about the end times, giving us a better understanding of biblical stories and the culture and religious practices of that time.

Chapter I.

In the first year after the Israelites left Egypt, on the sixteenth day of the third month, God spoke to Moses, saying, "Come up to Me on the mountain, and I will give you two stone tablets with My laws and commandments. You will teach them to the people."

Moses went up Mount Sinai, and God's glory covered the mountain in a cloud for six days. On the seventh day, God called to Moses from the cloud. His presence on the mountaintop looked like a blazing fire. Moses stayed on the mountain for forty days and forty nights, and during this time, God showed him past and future events, organizing all the laws and teachings.

God said, "Pay close attention to everything I am telling you and write it down in a book. In the future, people will realize that even when they sin and break My covenant, I have not abandoned them. When these events take place, they will understand that My judgments are right and fair. They will see that I have always been with them."

Write down everything I tell you today because I already know how stubborn and rebellious they will be. Even before I bring them into the land I promised to their ancestors—Abraham, Isaac, and Jacob—they will turn away from Me. They will enjoy all its blessings, eat until they are full, and then follow false gods that cannot save them when trouble comes. This will stand as a witness against them. They will forget My commandments and follow the sinful ways of the nations around them, worshiping idols and practicing evil. These false gods will become a trap and a burden to them.

Many will die or be taken captive by their enemies because they rejected My laws. They will stop celebrating My holy days, break My Sabbaths, and abandon the sacred place I gave them. Instead, they will build altars on high places, worship idols, and even sacrifice their children to demons. They will do terrible things because of the evil in their hearts.

I will send messengers to warn them, but they will refuse to listen. They will kill these messengers, persecute those who follow My law, and twist My words to justify their wrongdoing. Because of this, I will turn away from them and allow foreign nations to capture them and destroy their land. They will be scattered among different nations, and while in exile, they will forget My laws and commandments. They will lose understanding of My holy days and drift even further from Me.

But one day, they will return to Me with all their heart, soul, and strength. When they truly seek Me, they will find Me. I will bring them back from the nations where they were scattered and give them peace and righteousness. I will fill them with goodness and bless them instead of cursing them. They will no longer be the oppressed but the leaders.

I will place My sanctuary among them and live with them. I will be their God, and they will be My people, walking in truth and righteousness. I will never leave them because I am their Lord and God.

Moses fell on his face and prayed, saying, "O Lord, do not abandon Your people, Your chosen ones. Do not let them fall into the hands of their enemies, who will lead them further into sin. Show them mercy, Lord. Create a pure spirit within them so they do not continue down the path of evil and perish before You. They are Your people, whom You saved with great power from Egypt. Give them clean hearts and holy spirits so they do not fall into sin again."

The Lord answered Moses, "I know how stubborn they are. They will not fully obey Me until they admit their sins and the sins of their ancestors. But when they return to Me with all their heart and soul, I will change them. I will give them new hearts, and their children will follow Me as well. I will fill them with My holy spirit and purify them, so they will never turn away from Me again. They will obey My laws, and I will be their Father, and they will be My children. Everyone in heaven and on earth will know that they are My people, and I am their God. I love them with an everlasting love.

Write down everything I am telling you—past, present, and future. These words will stand for all generations, guiding them until I come to live among them forever."

Then God said to the angel of His presence, "Write everything down for Moses, from the beginning of creation until the time My sanctuary will be built forever among them." The Lord will reveal Himself to all, and everyone will know that He is the God of Israel, the Father of Jacob's descendants, and the eternal King who reigns from Mount Zion. Jerusalem and Zion will be holy forever.

The angel of the presence, who guided Israel through the wilderness, brought the tablets containing the history of the world— from the creation of time to the final renewal of heaven and earth. All creation will be restored as it was meant to be, and the Lord's sanctuary will be established in Jerusalem on Mount Zion. The stars and heavenly lights will be renewed, bringing healing, peace, and blessings to God's chosen people forever.

Chapter II.

Then the angel, following God's command, spoke to Moses and said, "Write down the full story of creation. Record how, in six days, God made everything and brought it to life. On the seventh day, He rested and made it a special, holy day for all time.

On the first day, God created the sky, the earth, and the waters. He also made angels—some to be in His presence, some to bring holiness, and others to control fire, wind, clouds, snow, hail, and frost. He created angels for thunder, lightning, and the changing seasons. He also made spirits for all His creatures, both in heaven and on earth. He formed the deep waters, darkness, evening, night, light, dawn, and daytime. Everything was made with His wisdom. We saw His creation and praised Him. Seven great things were made on the first day.

On the second day, God made the sky and placed it between the waters. Some waters rose above the sky, while others remained below, covering the earth. This was the only thing He created on the second day.

On the third day, God commanded the waters under the sky to come together so that dry land would appear. The waters obeyed, forming seas, rivers, and lakes. On this day, He also created dew, seeds, and plants. He made fruit trees, forests, and the Garden of Eden, filled with all kinds of plants. Four important things were made on the third day.

On the fourth day, God made the sun, moon, and stars. He placed them in the sky to shine on the earth, to separate day from night, and to mark time. These lights were also signs for the days, the Sabbath, the months, festivals, years, and special cycles of time. Three great things were made on the fourth day.

On the fifth day, God created the great sea creatures that live in the deep waters. These were the first living things He made. He also created fish and all creatures that live in water, as well as all kinds of birds. When the sun rose, it shone on these creatures and blessed them, along with all the plants and trees that grow on the earth. Three kinds of living beings were made on the fifth day.

On the sixth day, God made land animals, including livestock and creatures that move on the ground. After that, He created humans, making both man and woman. He gave them control over the earth, the seas, the birds, the animals, and all living things. They were put in charge of everything on earth. Four types of creation were made on the sixth day, bringing the total to twenty-two.

On this day, God finished all His work—the heavens, the earth, the seas, and everything in them. He established a special sign: the Sabbath. He commanded that people should work for six days and rest on the seventh.

God also told His angels to observe the Sabbath with Him, both in heaven and on earth. Then He said, "I will choose a special people from all the nations, and they will keep the Sabbath. I will make them My people and bless them, just as I have blessed and set apart this day for Myself. They will belong to Me, and I will be their God.

From everything I have seen, I have chosen Jacob's descendants as My firstborn son. I have set them apart forever and will teach them to honor the Sabbath, so they may rest and keep it holy."

That is why the Sabbath is a sign—a day to celebrate with food, drink, and praise to the Creator. Just as God chose a special people, they will keep the Sabbath and celebrate with us.

His commandments were given as a way to praise Him forever.

From Adam to Jacob, there were twenty-two generations, just like there were twenty-two kinds of work completed before the seventh day. The Sabbath was blessed along with the days before it, making it a time of holiness and rest.

To Jacob and his descendants, God gave the promise that they would be a holy and blessed people. This was part of His first law and covenant, just as He blessed the Sabbath.

In six days, God created the heavens, the earth, and everything in them. On the seventh day, He made it holy. He commanded that anyone who works on this day must be punished, and anyone who disrespects it will suffer.

Teach the Israelites to keep this day holy and rest from all work, for it is the most sacred of all days. Whoever disrespects it will be punished, and whoever works on it will face consequences forever. This law was given so that the Israelites would always observe the Sabbath and never lose their inheritance. It is a holy and blessed day.

Those who honor it and rest will also be holy and blessed, just as we are.

Tell the Israelites to always keep the Sabbath. Let them know that on this day, they should not do unnecessary work, seek their own pleasure, prepare food or drink, fetch water, or carry heavy loads through their gates. All of this must be done on the sixth day.

They should also not move things between houses on the Sabbath. This day is even more sacred than a jubilee. We in heaven have been observing the Sabbath long before it was given to humans.

The Creator blessed this day, but He did not require every nation to follow it. He set Israel apart to keep this law. Only they were chosen to eat, drink, and celebrate the Sabbath on earth.

The Creator made this day special, setting it apart as the holiest and most honored of all days.

This law was given to the Israelites as a lasting command for all generations.

Chapter III.

During the second week, over six days, we brought all kinds of animals to Adam. On the first day, he saw wild animals, on the second, livestock, on the third, birds, on the fourth, land creatures, and on the fifth, sea creatures.

Adam gave each one a name, and whatever he called them became their name. Over those days, he saw every kind of animal, both male and female, but he remained alone—there was no companion for him.

Then the Lord said to us, "It is not good for man to be alone. Let us make a helper for him."

So, God put Adam into a deep sleep. While he slept, God took one of his ribs and used it to create a woman. Then He closed the place where the rib had been removed.

When Adam woke up on the sixth day, God brought the woman to him. Adam immediately recognized her and said, "She is part of me—bone of my bones, flesh of my flesh. She will be called 'woman' because she was taken from man."

This is why a man leaves his parents and joins his wife, and the two become one.

Adam was created in the first week, and in the second week, the woman—made from his rib—was brought to him. God showed her to him, and because of this, a command was given: if a woman gives birth to a boy, she will be unclean for seven days, but if she gives birth to a girl, she will be unclean for fourteen days.

After Adam had lived in the land where he was created for forty days, we took him to the Garden of Eden so he could take care of it. His wife was brought into the Garden on the eightieth day, and from then on, they lived there together.

For this reason, a law was written on the heavenly tablets: "When a woman has a son, she will be unclean for seven days, just like the first week. Then, for thirty-three more days, she must stay away from anything holy and cannot enter the sacred place until her purification is complete."

If she has a daughter, she will be unclean for two weeks, just like the first two weeks, and will need sixty-six more days to complete her purification, making a total of eighty days.

Once these eighty days are over, she may enter the holy place again, because the Garden of Eden is holier than the rest of the earth, and every tree in it is sacred.

This is why the law was given regarding childbirth: a woman must not touch anything holy or enter the sacred place until her purification is complete.

This law was recorded for Israel so they would follow it for all generations.

During the first week of the first jubilee, Adam and his wife lived in the Garden of Eden for seven years. They took care of it, following the instructions they were given. Adam worked hard, and though he was naked, he felt no shame. He protected the garden from birds, animals, and livestock. He gathered fruit and saved some for himself and his wife.

After exactly seven years, in the second month, on the seventeenth day, the serpent approached the woman. It asked, "Did God really say you cannot eat from any tree in the garden?"

The woman replied, "We can eat from any tree except the one in the middle. God said, 'Do not eat from it or even touch it, or you will die.'"

The serpent said, "You won't die. God knows that if you eat it, your eyes will be opened, and you will be like gods, knowing good and evil."

The woman looked at the tree and saw that its fruit looked good to eat, was beautiful, and seemed to bring wisdom. She took some, ate it, and gave some to Adam. He ate as well.

Immediately, their eyes were opened, and they realized they were naked. They sewed fig leaves together to cover themselves.

God cursed the serpent, giving it eternal punishment. Then He turned to the woman and said, "Because you listened to the serpent and ate the fruit, I will greatly increase your pain in childbirth. You will suffer when you have children. You will long for your husband, and he will rule over you."

To Adam, He said, "Because you listened to your wife and ate from the tree I told you not to eat from, the ground is now cursed. You will have to work hard to get food from it for the rest of your life. It will produce thorns and weeds, and you will eat the plants of the field. You will sweat as you work for your food until the day you return to the ground. You were made from dust, and you will return to dust."

Then God made clothes from animal skins for Adam and his wife and dressed them. After that, He sent them out of the Garden of Eden.

On the day Adam left the garden, he offered a sacrifice at sunrise, burning fragrant spices like frankincense, galbanum, and stacte to seek forgiveness for his shame.

That same day, all animals, birds, and creatures that moved on the earth became silent. Before this, they had all spoken the same language. Then God sent all the creatures out of the Garden of Eden, separating them into their proper places. Of all the living things, only Adam was given clothing to cover his nakedness.

This is why it is written on the heavenly tablets that all who know the law's judgment must cover themselves and not expose their bodies as the other nations do.

On the new moon of the fourth month, Adam and his wife left the Garden of Eden and settled in the land of Elda, the place where they were created. Adam named his wife Eve.

During the first jubilee, they had no children. Later, Adam was with his wife, and he worked the land just as he had been taught in the Garden of Eden.

Chapter IV.

During the third week of the second jubilee, Eve gave birth to Cain. In the fourth week, she had Abel, and in the fifth week, she gave birth to their daughter, Âwân.

In the first year of the third jubilee, Cain killed Abel because God accepted Abel's offering but rejected his. Cain attacked Abel in a field, spilling his blood, which cried out to heaven for justice.

God confronted Cain about his crime, and as a result, Cain was cursed and became a wanderer. He had to live with the guilt of his brother's death. This is why it is written on the heavenly tablets: "Cursed is anyone who kills their neighbor in secret, and all who hear of it must say, 'So be it.' Those who stay silent share in the guilt."

This is why we confess all our sins before the Lord—whether they happen in heaven, on earth, in the open, or in secret—so that nothing remains hidden. Adam and Eve mourned Abel for many years. But in the fourth year of the fifth week, their sadness turned to joy when Adam was with his wife again, and she gave birth to Seth. Adam said, "God has given us another child to take Abel's place."

In the sixth week, Eve had a daughter named Azûrâ. Cain married his sister Âwân, and by the end of the fourth jubilee, she gave birth to their son, Enoch. In the first year of the first week of the fifth jubilee, people began building houses on the earth. Cain built a city and named it after his son, Enoch. Adam and Eve had nine more sons.

During the fifth week of the fifth jubilee, Seth married his sister Azûrâ, and in the fourth year of the sixth week, they had a son named Enos. Enos was the first to call on the name of the Lord.

In the third week of the seventh jubilee, Enos married his sister Nôâm, and in the third year of the fifth week, they had a son named Kenan. At the end of the eighth jubilee, Kenan married his sister Mûalêlêth, and in the third year of the first week of the ninth jubilee, they had a son named Mahalalel.

During the second week of the tenth jubilee, Mahalalel married Dinah, the daughter of Barakiel, who was his cousin. In the sixth year of the third week, they had a son named Jared.

During Jared's time, a group of angels called the Watchers came down to teach people about justice and judgment. In the eleventh jubilee, Jared married Baraka, the daughter of Râsûjâl, another relative. In the fifth week of that jubilee, she gave birth to Enoch.

Enoch was the first to learn writing, wisdom, and knowledge. He studied the signs in the sky, helping people understand time and seasons. He recorded the weeks, years, and Sabbaths exactly as they were revealed to him. He also received visions of past and future events, writing them down for future generations.

In the twelfth jubilee, during the seventh week, Enoch married Edna, the daughter of Danel, his cousin. In the sixth year of that week, they had a son named Methuselah.

Enoch spent six jubilees with the angels of God, learning about heaven and earth. He wrote everything down and warned against the Watchers, who had taken human wives. Because of his righteousness, God took Enoch away from the world and placed him in the Garden of Eden, where he recorded His judgments.

Later, God sent a flood to cleanse the earth because of human wickedness. Enoch's life stood as a warning to people. He also burned incense on a mountain, offering sweet-smelling spices to God. The Lord established four sacred places on earth: the Garden of Eden, the Mount of the East, Mount Sinai, and Mount Zion. These places would one day be made holy again to cleanse the world from its corruption.

In the fourteenth jubilee, Methuselah married Edna, the daughter of Azrial, his cousin. During the third week, in the first year of that week, she gave birth to Lamech.

In the fifteenth jubilee, Lamech married Betenos, the daughter of Baraki'il, his cousin. They had a son named Noah. Lamech said, "This child will bring us relief from the hardship caused by the cursed ground."

At the end of the nineteenth jubilee, during the seventh week of the sixth year, Adam died. His sons buried him in the land where he had been created. He lived 930 years but did not reach 1,000. According to the heavenly record, 1,000 years is like one day, so the prophecy about the tree of knowledge was fulfilled: "On the day you eat from it, you will die." In God's time, Adam died within that same "day."

That same year, Cain was killed when his house collapsed on him. This fulfilled what was written on the heavenly tablets: "Whoever kills with a weapon will be killed by the same." Since Cain killed Abel with a stone, he also died under stones as a just punishment.

In the twenty-fifth jubilee, Noah married Emzârâ, the daughter of Râkê'êl, his cousin. She gave birth to three sons: Shem in the third year, Ham in the fifth year, and Japheth in the first year of the sixth week.

Chapter V.

When the number of people on earth increased and they had daughters, the angels of God noticed how beautiful they were. During a certain year in a jubilee, they chose wives for themselves from among them, taking whoever they wanted. Their children grew into giants, and soon, lawlessness spread across the land. Corruption filled the world as humans, animals, and birds became violent, attacking and even eating each other. People's thoughts became completely evil, and wrongdoing took over everywhere.

God saw how wicked the world had become. All living creatures had strayed from His ways, and the earth was filled with sin. Because of this, He decided to destroy mankind and every living thing He had created. But Noah stood apart—he was righteous, and God looked upon him with favor.

God's anger also turned toward the angels He had sent to earth. He stripped them of their power and commanded that they be chained deep underground, cut off from the rest of creation. Their giant offspring were also condemned to die. God declared that they would be wiped out by the sword and removed from the earth. He said, "My Spirit will not remain with humans forever, for they are mortal. Their lifespan will now be 120 years." Then, He caused violence to spread among the people, making them turn against each other until they were completely destroyed.

The fallen angels, forced to watch their children die, were imprisoned in the depths of the earth. There they will remain until the

Translated by Tim Zengerink

final judgment when God will punish all those who corrupted the world. No one escaped judgment. The wicked were completely removed from the earth. Afterward, God gave a new, pure nature to all living creatures so they would not turn back to evil.

The heavenly tablets record the laws for all creation. It is written that anyone who strays from their rightful path will be judged. Every action—whether done in heaven, on earth, in darkness, in light, in the depths, or even in Sheol—is seen by God. He is a just judge who does not show favoritism or accept bribes. If He has decided on a judgment, He will carry it out completely.

However, God also promised that if Israel repents and turns back to Him, He will forgive their sins. Once a year, He will grant mercy to those who admit their guilt and change their ways. But those who became corrupt before the flood were given no mercy. Only Noah was accepted by God, and his righteousness saved not only himself but also his sons. He followed all of God's instructions exactly as he was told.

God decided to destroy everything on earth—humans, animals, and birds. He commanded Noah to build an ark to survive the flood. Noah obeyed, constructing the ark just as God had instructed. This happened in the twenty-seventh jubilee, during the fifth week, in the fifth year, on the new moon of the first month. In the sixth year, during the second month, Noah and the animals entered the ark. On the seventeenth evening, the Lord shut the door from the outside.

Then, God opened the seven floodgates of heaven and the seven deep springs of the earth. For forty days and nights, heavy rain poured down, and water gushed up from the ground. The floodwaters rose until they covered even the tallest mountains by fifteen cubits. The ark floated on the surface of the waters. For five months—150 days—the

66

flood remained over the earth. Finally, the ark came to rest on Mount Lubar, one of the mountains in the Ararat range.

In the fourth month, the deep springs of the earth were sealed shut, and the floodgates of heaven were closed. By the seventh month, the waters began to drain into the earth's depths. In the tenth month, the mountain peaks became visible again. By the first month of the new year, the land started to dry. On the seventeenth day of the second month, the ground was completely dry. Then, on the twenty-seventh day, Noah opened the ark, letting all the animals, birds, and creatures go free to return to the land.

Chapter VI.

On the first day of the third month, Noah left the ark and built an altar on the mountain. He made a sacrifice to cleanse the earth, taking a young goat and using its blood to remove the guilt of the land. Everything that once lived on the earth had been wiped out, except for those saved in the ark with Noah. He placed the fat of the sacrifice on the altar and also offered an ox, a goat, a sheep, young goats, salt, a turtledove, and a young pigeon. He poured oil over them, sprinkled wine, and burned frankincense, creating a pleasing aroma that rose to God.

The Lord smelled the offering and made a promise to Noah, vowing never to destroy the earth by a flood again. He said that planting and harvest, cold and heat, summer and winter, day and night would continue as they were meant to, without interruption. Then He told Noah, "Have many children and fill the earth. I have placed all animals on land, in the sea, and in the sky under your authority. Just as I gave you plants to eat, I now give you everything. But do not eat meat that still has blood in it, because life is in the blood. Anyone who takes another person's life will be held accountable, for humans were made in God's image. Be fruitful and spread across the earth."

Noah and his sons made a vow before God that they would never eat blood. This became a lasting agreement for all future generations. God commanded that Israel remember this covenant, just as He later instructed Moses on the mountain. They were to sprinkle blood as part of their worship and never eat blood from any animal, bird, or livestock. This law was given as a permanent reminder so that Israel would always keep this commandment. Anyone who ate blood would be cut off from the land, and their descendants would be forgotten before God.

To seal His promise, God gave Noah a sign—He placed a rainbow in the clouds as a symbol of His eternal covenant that He would never again flood the earth to destroy it. It was also recorded in the heavenly tablets that a special festival, the Feast of Weeks, would be celebrated every year in this month to renew the covenant. This festival had been observed in heaven from the creation of the world until Noah's time, lasting 26 jubilees and five weeks of years. Noah and his sons continued to celebrate it for seven jubilees and one week of years until Noah died. After that, people forgot about it until Abraham revived it. His son Isaac, then Jacob, and their descendants observed it until Moses renewed it on the mountain.

God commanded the people of Israel to keep this festival in every generation, celebrating it as a reminder of the covenant. It became known as both the Feast of Weeks and the Feast of Firstfruits, a special day written into the law. It was established that this festival would be celebrated on a specific day each year, and Moses was given instructions on how it should be observed, ensuring that the Israelites would keep it faithfully for generations.

The new moons of the first, fourth, seventh, and tenth months were also set as days of remembrance, marking the changing seasons and the divisions of the year. Noah established these days as lasting feasts. On the new moon of the first month, he was told to build the ark, and on that same day, the earth became dry after the flood. On the new moon of the fourth month, the deep springs of the earth were sealed shut. On the new moon of the seventh month, the earth's abysses opened, allowing the floodwaters to drain. On the new moon of the tenth month, the mountain peaks became visible, and Noah rejoiced. These days were recorded in the heavenly tablets and were meant to be remembered forever.

The year was divided into four seasons, each lasting 13 weeks, making a total of 52 weeks. This cycle was written and established in the heavenly tablets and was not to be changed. The Israelites were commanded to follow a 364-day year, so the feasts and seasons would remain in their correct order. Any changes to this system would throw the years out of sync, leading to confusion about the feasts, new moons, Sabbaths, and the seasons.

God warned Moses that after his death, the Israelites would eventually stray from this system. They would abandon the 364-day calendar, causing their festivals, holy days, and Sabbaths to fall out of order. Some would begin following the lunar calendar, leading to mistakes in their observances and making holy days unclean. They would mix the sacred with the ordinary, forget the commandments, and even start eating blood and all kinds of meat without following the laws.

This message was given as a warning, so the Israelites would not be led astray and so they would understand the serious consequences of ignoring the appointed times and the laws of the covenant.

Chapter VII.

In the seventh week of the first year of this special time, Noah planted vineyards on the mountain where the ark had landed—Mount Lubar, one of the Ararat Mountains. It took four years for the vines to grow and produce grapes. In the seventh month of that year, Noah gathered the grapes.

Noah made wine from the grapes and stored it in a container. He kept it until the fifth year, and on the first day of the first month, he held a big feast. As part of the celebration, he made a burnt offering to God, sacrificing a young ox, a ram, seven one-year-old sheep, and a young goat to ask for forgiveness for himself and his sons.

Noah started with the goat, using some of its blood and meat on the altar. He placed all the fat on the altar as part of the burnt offering. Then, he did the same with the ox, the ram, and the sheep, placing their meat and fat on the altar and mixing them with oil. He poured wine over the fire and burned incense. The smell rose up and pleased God.

Noah and his sons enjoyed the feast, drinking the wine with joy. As the evening came, Noah went to his tent, fell asleep, and became drunk. While he was asleep, he accidentally uncovered himself and lay naked.

His son Ham saw him like this and went to tell his brothers. But Shem and Japheth took a garment, placed it on their shoulders, and walked backward into the tent to cover their father without looking at him.

When Noah woke up and realized what had happened, he cursed Ham's son, Canaan, saying, "Canaan will be a servant to his brothers." Then he blessed Shem, saying, "Praise be to the Lord, the God of Shem. May Canaan serve him." He also prayed for Japheth, saying, "May God bless Japheth and let him live among Shem's people. And may Canaan be his servant too."

Ham was upset when he heard the curse on his son. He left Noah and went with his sons—Cush, Mizraim, Put, and Canaan—to build a city, naming it after his wife, Ne'elatama'uk.

Japheth, feeling jealous, built his own city and named it after his wife, 'Adataneses.

Shem, however, stayed close to Noah and built a city near him on the mountain, calling it Sedeqetelebab, after his wife.

This is how three cities were built near Mount Lubar: Sedeqetelebab in the east, Na'eltama'uk in the south, and 'Adataneses in the west.

The sons of Shem were Elam, Asshur, Arpachshad (who was born two years after the flood), Lud, and Aram. Japheth's sons were Gomer, Magog, Madai, Javan, Tubal, Meshech, and Tiras. These were the children and grandchildren of Noah.

In the twenty-eighth jubilee, Noah began teaching his grandsons the laws and commands he had received. He told his sons to live good and honest lives, to dress modestly, respect their Creator, honor their parents, love their neighbors, and avoid evil and impurity.

He warned them that the flood had come because of three great sins: wickedness, impurity, and the wrongdoing of the Watchers, who had taken wives from among human women.

These sins led to the birth of the Nephilim, who became violent and corrupt. The Giants killed the Nephilim, the Nephilim killed the Eljo, and the Eljo turned against humans. The world became filled with wickedness, violence, and bloodshed.

People didn't just harm one another; they also hurt animals, birds, and all living creatures. Blood covered the earth, and people's hearts became filled with evil thoughts and wicked plans.

Because of this, God wiped everything off the face of the earth. Every living thing was destroyed because of the violence and corruption.

Noah then spoke to his sons, saying, "I see what you are doing, and it worries me. You are not following the right path. You are becoming jealous of one another, arguing, and growing apart. I fear that after I die, you will start shedding blood, and because of this, you too will be removed from the earth.

Anyone who kills another person or drinks animal blood will be destroyed. Their family line will end, and they will have no descendants. Those who do such things will go to Sheol, a place of punishment and darkness, and they will be taken from the earth.

The killing of people and animals must stop. Any blood spilled must be covered because I have been commanded to warn you, your children, and all creatures.

Do not let your hearts be stained by the flesh of animals, for blood is life, and God will hold everyone accountable for the blood they shed. The earth cannot be cleansed of spilled blood unless the one who caused it also sheds their own blood. Only then will the land be pure.

Now, my children, listen to me. Live righteously and fairly, so that you will always walk in the right way. If you do good, your actions will rise before God, who saved me from the flood.

Go and build cities, plant trees, and grow crops. When you plant fruit trees, do not pick the fruit for the first three years. In the fourth year, the fruit will be holy, and the first portion should be given to God, the Creator of heaven and earth.

The remaining fruit will be for those who serve God. In the fifth year, you may harvest the fruit freely, and everything you plant will grow well.

This rule was first given by Enoch, the ancestor of your forefathers. Methuselah passed it down to his son Lamech, and Lamech taught me, just as their ancestors had taught them.

Now, I am passing this rule to you, just as Enoch gave it to his son long ago. He carefully taught his children and grandchildren, ensuring they followed it throughout their lives."

Chapter VIII.

In the first week of the twenty-ninth jubilee, at the very beginning, Arpachshad married Rasu'eja, the daughter of Susan, who was Elam's daughter. Three years later, they had a son named Kainam. As Kainam grew older, his father taught him how to write. One day, Kainam went out to look for a place to build a city. While exploring, he found an old inscription carved into a rock by earlier generations. He read what was written, copied it down, and unknowingly committed a sin. The inscription contained secret knowledge from the Watchers—beings who had studied the movements of the sun, moon, and stars to find hidden signs in the sky. Fearing Noah's anger, Kainam kept this discovery a secret.

In the first year of the second week of the thirtieth jubilee, Kainam married Melka, the daughter of Madai, who was Japheth's son. Four years later, they had a son named Shelah, who said, "I have truly been sent." In the fifth week of the thirty-first jubilee, Shelah married Mu'ak, the daughter of Kesed, his father's brother. Five years later, she gave birth to a son named Eber. Later, in the thirty-second jubilee, during the seventh week, Eber married 'Azûrâd, the daughter of Nebrod. In the sixth year of that week, they had a son named Peleg. He was given this name because, during his lifetime, Noah's descendants started dividing the land among themselves. They first made a secret agreement on how to split it, then told Noah about their decision.

At the start of the thirty-third jubilee, in the first year of the first week, the earth was officially divided into three regions: one for Shem, one for Ham, and one for Japheth. This division was made under the guidance of a messenger sent by God. Noah's descendants and their families gathered to decide the borders of each territory.

Shem's land was in the middle of the earth, and he and his descendants were meant to live there forever. His territory started at the middle of the Rafa mountains, where the Tina River begins. It stretched west along the river to the waters of the deep and flowed into the Sea of Me'at, then continued toward the Great Sea. Japheth's land was to the north of Shem's, while Ham's land lay to the south. Shem's territory extended south to Karaso and followed the western coastline of the Great Sea until it reached the mouth of the Egyptian Sea. From there, it went down the shores of the Great Sea to 'Afra, then reached the Gihon River. It followed the southern banks of the river eastward, passing just below the Garden of Eden. His land included Eden, the eastern regions, and all the way back to the Rafa mountains and the Tina River. This area, including the lands of Eden, was Shem's permanent inheritance.

Noah was pleased with this division because it matched the blessing he had spoken: "Blessed be the Lord God of Shem, and may the Lord dwell in his land." He knew that the Garden of Eden was the holiest place on earth, where God's presence was. He also knew that Mount Sinai, in the wilderness, and Mount Zion, at the center of the earth, were sacred places made to align with each other. Noah praised God for His wisdom and power and recognized the special blessing given to Shem and his descendants. Shem's land included Eden, the Red Sea region, India, the lands surrounding the Red Sea, and mountain ranges like Bashan, Lebanon, Sanir, Amana, Asshur, and Elam, as well as the Ararat region. This large territory was rich and full of resources.

Ham's land was located south of the Garden of Eden, beyond the Gihon River. His territory stretched to the fiery mountains and reached the 'Atel Sea. It extended west to the sea of Ma'uk, where many lost things eventually ended up. From there, it moved north to Gadir, followed the Great Sea's coastline, and circled back to the Gihon River, returning near the Garden of Eden. This land was given to Ham and his descendants as their permanent inheritance.

Japheth's land was north of the Tina River and covered the northern and northeastern regions, stretching all the way to the land of Gog. It extended eastward to the Qelt mountains and the sea of Ma'uk, then curved toward Gadir and beyond. To the west, it reached Fara, then turned toward 'Aferag before extending east to the Sea of Me'at. Japheth's land continued northeast, reaching the Tina River and the Rafa mountains. His territory was vast and included five large islands. Japheth's land was known for its cold climate, while Ham's land was hot, and Shem's land had a balanced, temperate climate in between.

Chapter IX.

Ham divided his land among his sons. Cush received the first portion, which was in the eastern region. To the west of Cush was Mizraim's land, followed by Put's territory even farther west. The farthest west, along the seacoast, was the land given to Canaan.

Shem also divided his inheritance among his sons. Elam received the first portion, which covered the land east of the Tigris River and stretched further east, including all of India. His territory also included the coastline along the Red Sea, the waters of Dedan, the mountains of Mebri and Ela, the land of Susan, and the regions near Pharnak, reaching the Red Sea and the Tina River.

The second portion was given to Asshur, which included all the land of Assyria, along with the cities of Nineveh and Shinar, as well as the borderlands of India. This region followed the path of the major rivers in the area.

The third portion went to Arpachshad, covering the entire land of the Chaldeans, east of the Euphrates River, near the Red Sea. His inheritance also included the desert lands near the sea's mouth facing Egypt, along with Lebanon, Sanir, and 'Amana, extending to the Euphrates River.

The fourth portion was given to Aram. His land was located between the Tigris and Euphrates Rivers, north of the Chaldean territory, and stretched to the Asshurite mountains and the land of 'Arara.

Lud, Shem's fifth son, received the mountains of Asshur and the surrounding lands. His territory extended to the Great Sea and stretched eastward, toward the land of his brother Asshur.

Japheth also distributed his land among his sons. Gomer received the first portion, which stretched eastward from the northern region to the Tina River. To the north, Magog's land included the inner northern territories, extending toward the Sea of Me'at.

Madai's land was located west of his brothers' territories and included islands and coastal regions.

Javan, who received the fourth portion, was given all the islands and lands that bordered Lud's territory.

Tubal received the fifth portion, which covered the central landmass bordering Lud. His land extended to the second landmass and stretched beyond into the third section of land.

Meshech was given the sixth portion, which extended beyond the third landmass and reached the eastern border of Gadir.

Tiras received the seventh portion, which included four large islands located in the sea. These islands extended toward the border of Ham's land. The Kamaturi Islands were also assigned to the descendants of Arpachshad as part of their inheritance.

Noah's sons divided the earth among their descendants while Noah was still alive. He made them take an oath, binding them under a curse so that no one would take land that was not given to them.

They all agreed and swore, saying, "So be it, so be it," for themselves and for their future generations. This agreement was meant to last forever, until the day of judgment. On that day, the Lord will judge all nations with fire and the sword for their sins, their corruption, and the evil they have spread across the earth.

Chapter X.

During the third week of that jubilee, unclean spirits began leading Noah's descendants astray, causing confusion and destruction. Noah's sons came to him, troubled by how these demons were deceiving, blinding, and even killing their children.

Noah prayed to the Lord, saying:

"God of all spirits, You have shown mercy to me and my family, saving us from the flood and sparing us from the fate of the wicked. Your kindness and compassion have been great.

Please continue Your grace upon my sons. Do not let these evil spirits take control of them or lead them to destruction. Bless me and my children so that we may grow in number and fill the earth. You know how the Watchers, the ancestors of these spirits, acted in my time. Bind these spirits and keep them in the place of punishment so they cannot harm Your faithful servants. They were created for destruction and should not have power over those who seek righteousness.

Do not let them rule over the living or the righteous, for only You have authority over all spirits. Do not allow them to overpower those who choose to walk in Your ways forever."

The Lord heard Noah's prayer and ordered us to bind the spirits. Then, Mastema, the leader of these spirits, stepped forward and said:

"Lord, Creator of all, allow some of them to remain with me to carry out my commands. Without them, I cannot fulfill my purpose of testing and corrupting mankind, for their wickedness is great."

The Lord replied:

"I will allow one-tenth of them to remain with you, but the other nine-tenths must be bound in the place of punishment."

The Lord then commanded one of us to teach Noah how to heal diseases, knowing that mankind would continue to act wickedly. We obeyed, binding the evil spirits and leaving only one-tenth under Mastema's control on earth. We also taught Noah how to use herbs from the earth to heal sickness and resist the temptations brought by these spirits. Noah wrote everything down in a book, recording all the

78

remedies and instructions we gave him. This knowledge helped protect his descendants from harm.

Noah passed these teachings to his eldest son, Shem, whom he loved most. After living a righteous life, Noah died and was buried on Mount Lubar in the land of Ararat. He lived for 950 years, completing 19 jubilees, two weeks, and five years. Among all men, he was the most righteous, second only to Enoch, who served as a testimony for future generations until the day of judgment.

During the thirty-third jubilee, in the first year of the second week, Peleg married Lomna, the daughter of Sina'ar. In the fourth year, she gave birth to a son, Reu. Peleg said, "Look, the people have become wicked, for they have begun to build a city and a tower in the land of Shinar."

The people had moved away from the land of Ararat and settled in Shinar. There, they decided to build a great city and a tower that would reach the heavens. They said, "Let's build a tower so high that we will make a name for ourselves." They worked for 43 years, making bricks and using asphalt from the sea as mortar. The tower rose to an incredible height of 5,433 cubits and 2 palms, with its walls extending 13 and 30 stades.

The Lord said, "They are one people with a single language, and now nothing they plan will be impossible for them. Let us go down and confuse their language so they will not understand one another."

We went down with the Lord to see the city and the tower. He confused their language, making them unable to continue their work. The land of Shinar was called Babel because the Lord scattered the people across the earth. A powerful wind knocked the tower down, and its ruins remained between Asshur and Babylon in Shinar. The people were dispersed in the first year of the thirty-fourth jubilee.

Ham and his sons moved to their assigned land in the south. However, Canaan saw the land of Lebanon, from the river of Egypt, and found it desirable. Instead of settling in the land given to him by the sea, he chose to stay in Lebanon, east and west of the Jordan and along the coast.

Ham, along with Cush and Mizraim, warned Canaan:

"You have taken land that does not belong to you. This land was given to Shem and his descendants. If you stay here, you and your children will be cursed and driven out for your disobedience. Do not live in Shem's territory, for it was assigned to him and his descendants by God."

But Canaan ignored them and remained in Lebanon, from Hamath to the borders of Egypt, along with his sons. That is why the land became known as Canaan.

Japheth and his sons moved to their rightful land by the sea. However, Madai did not like his assigned territory. He asked Ham, Asshur, and Arpachshad for a piece of their land. He settled in Media, near his brother-in-law, and named the land after himself. The name Media has remained ever since.

Chapter XI.

During the thirty-fifth jubilee, in the third week of its first year, Reu married a woman named Ôrâ, the daughter of Ûr, who was the son of Kesed. In the seventh year of that week, she gave birth to a son named Serôh. Around this time, Noah's descendants started fighting among themselves. They captured and killed each other, spilling human blood across the land. Some even began eating blood. They built fortified cities, walls, and towers. People became proud, forming the first kingdoms and waging wars—city against city, nation against nation.

Weapons were created, children were trained in warfare, and men started capturing others to sell as slaves.

Ûr, the son of Kesed, built the city of Ara in the land of the Chaldees, naming it after himself and his father. The people made molten idols and began worshiping them. They carved statues and created impure images. Evil spirits led them further into sin, and Mastêmâ, the ruler of these spirits, worked tirelessly to spread corruption. He sent his demons to commit all kinds of wickedness, violence, and bloodshed. Because of the widespread sin during this time, Serôh was later called Serug, a name reflecting the transgressions of that era.

Serug grew up in Ur of the Chaldees, near his wife's mother's family. Sadly, he also worshiped idols. During the thirty-sixth jubilee, in the first year of the fifth week, Serug married Melka, the daughter of Kaber, who was the daughter of his father's brother. In the first year of that week, she gave birth to a son named Nahor. Nahor was raised in Ur of the Chaldees, where his father taught him the ways of their people, including astrology and fortune-telling.

During the thirty-seventh jubilee, in the first year of the sixth week, Nahor married 'Ijaska, the daughter of Nestag from the Chaldees. In the seventh year of that week, they had a son named Terah. During this time, Mastêmâ sent flocks of ravens and birds to destroy crops and steal seeds before they could be planted. As a result, Terah's name was given to symbolize the hardship caused by the ravens. The land became barren, and people struggled to gather enough food from their harvests.

In the thirty-ninth jubilee, during the first year of the second week, Terah married Edna, the daughter of Abram, who was his father's sister. In the seventh year of that week, she gave birth to a son, whom they named Abram, after his grandfather, who had died before his birth.

As Abram grew older, he began to see the mistakes and wickedness of the world. He noticed that the people around him worshiped idols and practiced unclean rituals. His father taught him how to write, but by the time he was 14, Abram chose to distance himself from his father's ways to avoid idol worship. Instead, he prayed to the Creator of all things, asking for guidance to live a pure and righteous life.

When the season for planting arrived, people gathered in the fields to guard their seeds from the ravens. Although he was still a boy, Abram went with them. When a massive flock of ravens came to eat the seeds, Abram ran toward them and shouted, "Do not come down! Go back to where you came from!" Miraculously, the ravens turned away. That day, Abram drove the ravens away seventy times, and not a single bird remained in the land where he lived.

The people were amazed and word of Abram spread throughout the Chaldees. That year, farmers sought his help, and he guided them during the planting season. With his assistance, they successfully sowed all their seeds. That year, the harvest was abundant, and the people rejoiced.

In the first year of the fifth week, Abram invented a new tool for oxen plows to prevent ravens from stealing the seeds. He designed a wooden container that attached to the plow, allowing seeds to drop directly into the soil as the oxen moved. This clever invention prevented birds from eating the seeds before they could grow. The people quickly adopted Abram's design, and soon, they were able to plant and harvest without fear of the birds. Abram's wisdom and leadership brought prosperity and peace to the land.

Chapter XII.

During the sixth week, in the seventh year, Abram spoke to his father, Terah, and said, "Father!"

Terah answered, "I am here, my son."

Abram asked, "Why do you worship and bow down to these idols? They have no life, no spirit, and they only lead people astray. Why put your trust in them?

Worship the God of heaven, the One who sends rain and dew, who controls everything on earth, and who created all things with His word. All life comes from Him.

Why believe in lifeless statues made by human hands? You carry them on your shoulders, yet they cannot help you. They bring shame to those who make them and deceive those who worship them. Do not follow them."

Terah replied, "I know what you are saying is true, my son, but what can I do? The people here force me to serve these idols. If I speak against them, they will kill me because they are committed to worshiping them. Stay silent, my son, or they will kill you too."

Abram shared these thoughts with his brothers, but they became angry with him. So he remained quiet.

During the fortieth jubilee, in the second week and the seventh year, Abram married Sarai, his father's daughter, and she became his wife. His brother Haran also married in the third year of the third week, and in the seventh year, his wife gave birth to a son named Lot. Their other brother, Nahor, also took a wife.

When Abram was sixty years old, in the fourth week and the fourth year, he woke up in the middle of the night and set fire to the house of idols, burning everything inside. No one knew he had done it. When the people awoke, they rushed to save their gods from the fire. Haran tried to rescue them, but the flames overtook him, and he died in Ur of the Chaldees in front of his father, Terah. They buried him there.

After this, Terah left Ur with his sons and traveled toward the land of Lebanon and Canaan. He settled in Haran, where Abram stayed with him for fourteen years.

During the sixth week, in the fifth year, Abram stayed awake on the new moon of the seventh month to observe the stars and predict the coming rainfall. As he watched, he thought, "All the movements of the stars, the moon, and the sun are in the hands of the Lord. Why am I searching for answers in them?

If God wills, He sends rain in the morning and evening. If He chooses, He withholds it. Everything happens according to His will."

That night, Abram prayed, saying, "My God, the Most High, You are my only God, and I have chosen to follow You alone. You created everything, and all that exists is the work of Your hands.

Protect me from the evil spirits that mislead people so that I will not be led away from You. Strengthen me and my descendants forever, so we will never turn from Your ways."

Then he asked, "Should I return to Ur of the Chaldees, where they are calling me back? Or should I stay here? Guide me, O God, and make my path clear so that I do not follow my own desires."

After Abram finished praying, the word of the Lord came to him, saying:

"Leave your land, your people, and your father's house, and go to the land I will show you. I will make you into a great and mighty nation.

I will bless you and make your name great, and through you, all the families of the earth will be blessed. I will bless those who bless you and curse those who curse you.

Do not be afraid, for I will be your God forever, for you and your descendants through all generations."

The Lord also commanded, "Open his mouth and ears so that he may understand and speak the language I have given." This was the original language of creation, which had not been spoken since the Tower of Babel. At that moment, Abram's mouth, ears, and lips were opened, and the Lord spoke to him in Hebrew.

Abram then took the writings of his ancestors, which were written in Hebrew, and copied them, studying them carefully. The Lord helped him understand what he could not comprehend, and Abram spent six months of the year studying these writings during the rainy season.

In the seventh year of the sixth week, Abram told his father that he planned to leave Haran to visit the land of Canaan and return later.

Terah said to him, "Go in peace. May the eternal God guide you, protect you from harm, and grant you grace, mercy, and favor in the eyes of those you meet. May no one have the power to hurt you. Go safely.

If you find a land that pleases you, take me with you, and take Lot, the son of your brother Haran, as your own. But leave Nahor, your brother, with me. When you return safely, we will all go with you."

Chapter X III.

Abram left Haran with his wife Sarai and his nephew Lot, traveling to the land of Canaan. On the way, they passed through Asshur and arrived at Shechem, where they settled near a large oak tree. The land was beautiful, stretching from Hamath to the great oak. Then, the Lord appeared to Abram and said, "I will give this land to you and your future family." Abram built an altar there and made a burnt offering to God.

After that, Abram moved to a mountain between Bethel in the west and Ai in the east, where he set up his tent. He saw that the land was rich and full of life, with vineyards, fig trees, pomegranates, olive trees, cedars, date palms, and many other plants. Water flowed from the mountains, making the land fertile. Abram gave thanks to the Lord, who had guided him safely from Ur of the Chaldees to this land of blessings.

In the first year of the seventh week, on the first day of the first month, Abram built an altar on the mountain and prayed, saying, "You are the eternal God, and You are my God." He offered a burnt sacrifice, asking God to always be with him. Then, Abram traveled south and arrived in Hebron, where a city was being built. He stayed there for two years before moving further south to Bealoth. While he was there, a famine spread across the land.

In the third year of that time, Abram went to Egypt and stayed for five years. During his stay, Pharaoh took Sarai into his palace. But the Lord sent terrible plagues upon Pharaoh and his household because of her. As a result, Abram became very wealthy, gaining many sheep, cattle, donkeys, horses, camels, servants, silver, and gold. His nephew Lot also became rich.

Pharaoh returned Sarai to Abram and sent them out of Egypt. Abram traveled back to the place where he had first set up his tent, between Bethel and Ai, near the altar he had built before. There, he thanked the Lord for bringing him back safely. In the forty-first jubilee, during the third year of the first week, Abram made another burnt offering at the altar and prayed, saying, "You are the Most High God, and You are my God forever."

In the fourth year of that time, Lot separated from Abram and moved to Sodom, where the people were extremely sinful. Abram was saddened by Lot's choice, especially since he had no children of his own. Later that year, after Lot was taken captive, the Lord spoke to Abram and said, "Look around in every direction—north, south, east, and west. I will give all this land to you and your descendants forever. Your family will be as numerous as the dust of the earth. Walk through the land, for it will belong to you and your future generations."

Abram then moved to Hebron and settled there. That same year, Chedorlaomer, the king of Elam, joined forces with Amraphel, the king of Shinar, Arioch, the king of Sellasar, and Tergal, the king of the nations. Together, they attacked the king of Gomorrah. The king of Sodom fled, and many people were wounded in the Siddim Valley near the Salt Sea. The invading kings took over Sodom, Adam, and Zeboim, capturing Lot and taking all his belongings to Dan.

One of the survivors escaped and told Abram what had happened to Lot. Abram quickly gathered his trained servants, armed them, and went after the enemy. He successfully rescued Lot, recovered his possessions, and brought back the people who had been taken. After the victory, Abram gave one-tenth of the recovered goods to the Lord, establishing a lasting rule that a tenth of all produce—grain, wine, oil, cattle, and sheep—should be given to the priests who serve before God.

The king of Sodom then approached Abram, bowed before him, and said, "Lord Abram, keep the goods for yourself, but return the people you rescued to me." Abram replied, "I swear to the Most High God that I will take nothing from you—not even a thread or a sandal strap—so that you cannot say, 'I made Abram rich.' The only things I will take are what my men have already eaten and the share that belongs to Aner, Eschol, and Mamre, who helped me. They will receive their portion."

Chapter XIV.

In the fourth year of that time, on the first day of the third month, the Lord spoke to Abram in a vision, saying, "Do not be afraid, Abram. I am your protector, and your reward will be very great." Abram replied, "Lord, what can You give me if I still have no children? The one who will inherit everything I own is Eliezer of Damascus, the son of my servant Maseq. You have not given me any children of my own."

The Lord answered, "Eliezer will not be your heir. You will have a child of your own, and he will inherit everything." Then, the Lord took Abram outside and said, "Look up at the sky and try to count the stars, if you can." As Abram looked at the endless stars above him, the Lord said, "That is how many descendants you will have." Abram believed what the Lord had promised, and because of his faith, God considered him righteous. Then the Lord said, "I am the one who brought you out of Ur of the Chaldees to give you this land as your inheritance forever. I will be your God and the God of your descendants."

Abram asked, "Lord, how can I be sure that I will inherit this land?" The Lord told him, "Bring Me a three-year-old cow, a three-year-old goat, a three-year-old ram, a turtledove, and a young pigeon." Abram did as the Lord commanded. In the middle of the month, he stayed near the oak trees of Mamre, close to Hebron. There, he built an altar, sacrificed the animals, and poured their blood on it. He cut the animals in half and placed the pieces across from each other, but he did not divide the birds. Then, large birds came down, trying to eat the sacrifices, but Abram chased them away.

As the sun began to set, Abram fell into a deep sleep. A heavy darkness surrounded him, filling him with fear. Then, the Lord spoke, saying, "Know for certain that your descendants will live in a land that is not their own. They will be enslaved and mistreated for 400 years. But I will punish the nation that enslaves them, and in the end, they will leave with many possessions. You, however, will live in peace and grow old before being buried. In the fourth generation, your descendants will return to this land, for the sins of the Amorites are not yet complete."

When Abram woke up, the sun had already set. He saw a blazing fire and a cloud of smoke pass between the divided pieces of the sacrifice. That day, the Lord made a covenant with Abram, saying, "I will give this land to your descendants—from the river of Egypt to the great river, the Euphrates. This land belongs to the Kenites, Kenizzites, Kadmonites, Hittites, Perizzites, Rephaim, Amorites, Canaanites, Girgashites, and Jebusites."

Abram completed the offerings, including the birds and their grain and drink offerings, which were burned in the fire. The Lord sealed His covenant with Abram on that day, just as He had done with Noah in this same month. Abram renewed the festival and the practice as a tradition for himself and his future generations.

Overjoyed, Abram shared everything that had happened with his wife, Sarai. He believed fully in God's promise that he would have many descendants. However, Sarai still had not been able to conceive. She told Abram, "Take my Egyptian maid, Hagar, as your wife. Maybe I can have children through her." Abram listened to Sarai and agreed. She gave Hagar to Abram as his wife. He was with Hagar, and she became pregnant and gave birth to a son. Abram named him Ishmael. This happened in the fifth year of that time, when Abram was 86 years old.

Chapter XV.

In the fifth year of the fourth week of this special time, during the third month, Abraham celebrated the Festival of First Fruits from the grain harvest. He made offerings to God on the altar, presenting the first portion of his crops. He sacrificed a young cow, a goat, and a sheep as burnt offerings, along with grain and drink offerings, and sprinkled frankincense on the altar.

The Lord appeared to Abraham and said, "I am God Almighty. Follow My ways and live with honesty and integrity. I will make a covenant with you and greatly increase your descendants." Abraham bowed down to the ground, and God continued, "My covenant is with you, and you will become the ancestor of many nations. From now on, your name will no longer be Abram but Abraham, because I have made you the father of many nations. I will bless you abundantly, and your family will grow. Nations and kings will come from you. This covenant will last forever between Me and your descendants. I will be your God and the God of your future generations. I will give you and your family the land of Canaan, where you now live as foreigners, as a permanent possession, and I will be their God."

Then God gave Abraham instructions, saying, "You and your descendants must keep My covenant for all generations. Every male among you must be circumcised as a sign of this agreement. On the eighth day after birth, every male must be circumcised, whether he was born in your household or bought from a foreigner. This will be a permanent mark of the covenant between us. Any male who is not circumcised on the eighth day has broken My covenant and will be cut off from his people."

God also said, "As for your wife, she will no longer be called Sarai. Her name will now be Sarah. I will bless her, and she will give birth to a son. She will be the mother of nations, and kings will come from her." Abraham bowed down and laughed to himself, thinking, "How can a hundred-year-old man have a child? Can Sarah, at ninety years old, give birth?" Then Abraham said to God, "If only Ishmael could receive Your blessing!"

But God replied, "No, Sarah will give you a son, and you will name him Isaac. My everlasting covenant will be with him and his descendants. As for Ishmael, I have heard your request. I will bless him, make him fruitful, and give him many descendants. He will become the father of twelve rulers, and I will make him into a great nation. But My covenant will be with Isaac, whom Sarah will give birth to at this time next year."

After God finished speaking, He left. Abraham immediately obeyed God's command. That same day, he circumcised Ishmael, every male in his household—whether born there or bought—and himself. Every male in his household was circumcised that day as a sign of the covenant. The commandment to circumcise boys on the eighth day was recorded as a permanent law on the heavenly tablets. Anyone who failed to do this would be removed from the covenant and separated from God's people.

All the angels who serve in God's presence were created to worship Him, and God chose Israel as His special people even before they existed. He declared that Israel would always be with Him and His holy angels. The children of Israel were commanded to keep this covenant forever. If they remained faithful, they would never be removed from their land. This law was given as an everlasting command.

Even though Ishmael and Esau were Abraham's descendants, God did not choose them to be near Him. He chose Israel, set them apart, and made them His people, different from all other nations. While every nation belongs to God, He placed spiritual rulers over them for guidance. But He reserved Israel for Himself, without any intermediaries. He alone is their God, leading and protecting them forever.

However, I must warn you that the children of Israel will not keep this covenant. They will fail to circumcise their sons, ignoring this eternal law, and will leave them uncircumcised as they were born. This will anger the Lord because they will have abandoned His covenant, rejected His laws, and followed the customs of other nations. Their rebellion and disrespect will lead to their exile, and they will be cast out of their land. There will be no forgiveness for them because they will have broken this everlasting covenant.

Chapter XVI.

At the start of the fourth month, under the shade of the large oak at Mamre, we visited Abraham. We told him that his wife, Sarah, would have a son. Sarah, listening nearby, laughed quietly to herself because she didn't believe it was possible. We reassured her and told her not to be afraid, though she denied laughing. Still, we revealed that her son's name, Isaac, had already been written in the heavenly records. We promised that when we returned at the right time, she would be expecting a child.

During that same time, the Lord brought judgment upon Sodom, Gomorrah, Zeboim, and the surrounding areas near the Jordan. Fire and sulfur rained down, destroying them completely. To this day, those cities remain in ruins. Their wickedness had reached its limit, as they had corrupted themselves and spread evil everywhere. Just as Sodom was punished, so will any place that follows in their footsteps.

However, God showed mercy to Lot because of Abraham. He rescued Lot from the destruction, but even after escaping, Lot and his daughters committed a terrible sin—something unheard of since the time of Adam. Their actions were recorded as a serious wrongdoing in the heavenly records. Because of this, it was decided that Lot's descendants would not survive. Just like Sodom, his family line would be cut off. Their judgment is certain, and when the time comes, none of them will remain.

That same month, Abraham left Hebron and traveled toward the area between Kadesh and Shur. He settled in the mountains near Gerar. By the middle of the fifth month, he moved again, this time to the Well of the Oath. Then, in the middle of the sixth month, as promised, the Lord visited Sarah, and she became pregnant. Just as God had said, Sarah gave birth to a son in the third month, on the day of the Festival of First Fruits of the Harvest. And so, Isaac was born, fulfilling God's promise.

We told Abraham that while his other sons would be connected to different nations, Isaac's descendants would be set apart as a holy people, chosen by God. His family line would belong to the Most High, forming a special kingdom and priesthood devoted to serving Him. After delivering this message, we left Abraham and went to Sarah, sharing the same words with her. Both Abraham and Sarah rejoiced deeply at this news.

Abraham built an altar to honor the Lord, who had protected him and blessed him with great joy, even though he was living in a foreign land. At the altar near the Well of the Oath, he held a seven-day festival of celebration. During this time, he made temporary shelters for himself and his household. This was the first time the Feast of Tabernacles was observed on earth. Every day for seven days, Abraham made offerings to the Lord, including two oxen, two rams, seven sheep, and one male goat as a sin offering, asking for forgiveness for himself and his descendants.

Along with these offerings, he also gave thanksgiving sacrifices, which included seven rams, seven goats, seven sheep, and seven male goats, as well as grain and drink offerings. He burned all the fat on the altar as a pleasing aroma to the Lord. Every morning and evening, Abraham burned incense made from a special blend of spices: frankincense, galbanum, stacte, nard, myrrh, costus, and other fragrant spices. He combined them in equal amounts to create a pure and sweet-smelling incense for God.

For the full seven days, Abraham joyfully celebrated the festival with complete devotion. His entire household took part in the observance, but no outsiders or uncircumcised people were allowed to join. Abraham praised God, giving thanks for creating him and guiding him according to His divine plan. God already knew that Abraham's descendants would follow the path of righteousness, and from his family would come a holy people who reflected His goodness.

With joy and respect, Abraham honored God and named this celebration the "Festival of the Lord," a time of rejoicing that pleased the Most High. We blessed Abraham and his descendants forever because he followed the festival exactly as it was written in the heavenly records. Because of this, it was decided in the heavenly writings that Israel would celebrate the Feast of Tabernacles every year for seven days in the seventh month. This was to be a lasting commandment for all generations.

This festival would never be forgotten. It was established that Israel must observe it every year. They were instructed to live in temporary shelters, wear wreaths on their heads, and take leafy branches and willow branches from the streams. Abraham gathered palm branches and beautiful fruits, and each morning, he walked around the altar seven times, giving thanks and praising God with great joy for all that He had done.

Chapter XVII.

In the first year of the fifth week of that special time, Isaac was weaned, and Abraham held a great feast in the third month to celebrate. Ishmael, the son of Hagar the Egyptian, stood beside Abraham, and Abraham felt great joy. He praised God for giving him sons and not leaving him without children. He also remembered the promise God had made to him when Lot had separated from him, and his heart was full of gratitude as he gave thanks to the Creator.

However, Sarah saw Ishmael playing and dancing while Abraham was celebrating, and she became jealous. She said to Abraham, "Get rid of this slave woman and her son. Her son will not share in the inheritance with my son, Isaac." Abraham was deeply troubled by this because it involved both his servant and his son.

But God spoke to Abraham and said, "Do not be distressed about the boy or the maidservant. Listen to what Sarah is saying, because your descendants will come through Isaac. But do not worry—I will also make a great nation from the son of the slave woman because he is your child too."

The next morning, Abraham got up early, took some bread and a skin of water, placed them on Hagar's shoulder along with her son, and sent them away.

Hagar wandered in the wilderness of Beersheba. When the water ran out, the child became weak and collapsed. She placed him under the shade of an olive tree and walked a short distance away, saying, "I cannot bear to watch my child die." She sat down and wept.

Then an angel of God appeared to her and said, "Hagar, why are you crying? Get up, take the child, and hold him, for God has heard your cries and seen your child's suffering." Suddenly, Hagar saw a well of water. She quickly filled the water skin and gave the boy a drink. Then, they continued on to the wilderness of Paran. The boy grew up and became an excellent archer, and God was with him. Later, Hagar found him a wife from Egypt, and she gave birth to a son. He named him Nebaioth, saying, "The Lord was near to me when I called upon Him."

In the first year of the seventh week, on the twelfth day of the first month, voices from heaven spoke about Abraham. They declared that he had been faithful in everything God had commanded him, that he truly loved the Lord, and that he had proven his loyalty in every test.

Then Mastêmâ, the adversary, came before God and said, "Abraham may love You, but he loves his son Isaac even more. Command him to offer Isaac as a burnt sacrifice on the altar, and then You will see if he is truly obedient. Then You will know if he is faithful in everything."

But the Lord already knew Abraham's heart and that he was strong through every test. God had already tested him when He called him to leave his homeland, when he faced famine, when he encountered the riches of kings, and when his wife was taken from him. God tested him when He gave him the covenant of circumcision and again when he had to send Ishmael and Hagar away. Through all these trials, Abraham remained faithful and patient. He never hesitated to follow God's instructions because he loved the Lord and was completely devoted to Him.

Chapter XVIII.

One day, God called out, "Abraham, Abraham!" and Abraham answered, "Here I am." Then God said, "Take your son, Isaac—the one you love—and go to the high mountains. There, on a mountain that I will show you, offer him as a burnt sacrifice."

Early the next morning, Abraham got up, saddled his donkey, and took two of his servants along with Isaac. He cut the wood for the burnt offering and set out for the place God had told him about. After traveling for three days, he saw the mountain in the distance.

When they arrived at a well, Abraham said to his servants, "Stay here with the donkey. Isaac and I will go up the mountain to worship. After we have worshiped, we will come back to you."

Abraham took the wood for the offering and placed it on Isaac's shoulders. He himself carried the fire and the knife as they walked together toward the mountain.

On the way, Isaac spoke to his father, saying, "Father?" Abraham replied, "Yes, my son?" Isaac asked, "We have the fire and the wood, but where is the lamb for the burnt offering?"

Abraham answered, "God will provide the lamb for the sacrifice, my son." And the two of them continued on.

When they reached the place God had chosen, Abraham built an altar and arranged the wood on it. Then, he tied up Isaac and placed him on the altar, on top of the wood. Abraham reached for the knife and was about to sacrifice his son.

At that moment, I was there, along with Mastêmâ, and we heard God call out, "Abraham, Abraham!" Abraham quickly responded, "Here I am." God said, "Do not harm the boy. Now I know that you truly respect and obey Me, because you were willing to offer your son, your only son."

Then, I called out to Abraham from heaven again, saying, "Abraham, Abraham!" He answered, "Here I am." I told him, "Do not lay a hand on the boy. You have shown your deep trust in God by not holding back your son from Him."

Mastêmâ was left in shame. Abraham looked up and saw a ram caught by its horns in the bushes. He went over, took the ram, and offered it as a sacrifice in place of Isaac.

Abraham named that place "The Lord Will Provide," and even today, people say, "On the mountain of the Lord, it will be provided."

Then God called out to Abraham again from heaven and said, "I swear by Myself," declares the Lord, "Because you have done this and did not hold back your beloved son, I will bless you greatly. I will make your descendants as numerous as the stars in the sky and the grains of sand on the shore. They will conquer the cities of their enemies, and through your descendants, all the nations of the earth will be blessed, because you have obeyed My voice. You have proven your faithfulness in everything I have asked of you. Now go in peace."

Abraham returned to his servants, and together they traveled back to Beersheba, where he settled near the Well of the Oath.

From that time on, Abraham celebrated this event every year for seven days with great joy. He named it the Festival of the Lord, remembering the seven days of his journey and safe return.

It is recorded in the heavenly writings that Israel and its future generations must observe this festival every year for seven days, celebrating with joy.

Chapter XIX.

In the first year of the first week of the forty-second jubilee, Abraham returned and settled near Hebron, in a place called Kirjath Arba. He lived there for fourteen years.

In the first year of the third week, Sarah passed away in Hebron. Abraham mourned for her and arranged her burial. During this time, he was tested to see if he would remain patient and free from anger, and he passed the test by staying calm and composed.

He kindly approached the sons of Heth and asked for a burial place for his wife. The Lord made them favor him, and they treated him with respect. Abraham asked for the field that contained the cave near Mamre, also known as Hebron. They agreed to give it to him for four hundred pieces of silver. Even though they offered it as a gift, Abraham insisted on paying the full price. After completing the purchase, he bowed before them twice and buried Sarah in the cave.

Sarah lived for 127 years—two jubilees, four weeks, and one year in total. Her passing was Abraham's tenth test of faith, and he remained faithful and patient. Even though God had promised him the land, he still humbly asked for a burial site instead of questioning God's promise. Because of his faith, he was honored in the heavenly records as a friend of God.

In the fourth year of that time, Abraham arranged a marriage for Isaac. He chose Rebecca, the daughter of Bethuel, who was the son of Nahor, Abraham's brother. Around the same time, Abraham also married another wife, Keturah, who came from the daughters of his servants. Hagar had already passed away before Sarah. Over the next fourteen years, Keturah gave birth to six sons: Zimram, Jokshan, Medan, Midian, Ishbak, and Shuah.

In the second year of the sixth week, Rebecca gave birth to twin sons, Jacob and Esau. Jacob was quiet and righteous, living peacefully in tents, while Esau was wild, spending his time hunting in the fields and becoming skilled in battle. Abraham loved Jacob, but Isaac favored Esau.

As Abraham watched Esau's behavior, he realized that Jacob, not Esau, would carry on his name and legacy. He called Rebecca, knowing she loved Jacob more than Esau, and said:

"My daughter, take great care of Jacob,
  For he will inherit my place on this earth.
He will bring blessings to all people
  And will bring honor to the line of Shem.
The Lord has chosen him to be His own people,
  Set apart from all the nations of the earth.
Though Isaac loves Esau more,
  You love Jacob, and I ask you to care for him even more.
Let your love for him guide your actions,
  For he will bring blessings to us
  And to all future generations.
Be strong and take joy in your son Jacob,
  For I have loved him more than all my children.
He is blessed forever,
  And his descendants will fill the earth.
If a man could count the grains of sand on the earth,
  Jacob's descendants would be just as many.
All the blessings God has given me
  Will belong to Jacob and his descendants forever.
Through his family, my name and the names of my ancestors—
  Shem, Noah, Enoch, Mahalalel, Enos, Seth, and Adam—will
  be honored.

These blessings will uphold the heavens,
  Strengthen the earth,
  And renew the stars above."

Then Abraham called Jacob to stand before Rebecca and kissed him. He blessed him, saying:

"My beloved son Jacob, whom I cherish,
  May God bless you from the heavens above.
May He give you all the blessings He gave to Adam, Enoch,
  Noah, and Shem.
May all the promises He made to me and our family be fulfilled in
  you,
  And may those blessings last forever, as long as the heavens
  remain above the earth.
May no spirit of Mastêmâ have power over you or your
  descendants,
  To lead you away from the Lord your God,
  From this day forward and forever.
May the Lord be your Father,
  And may you be His firstborn son,
  A blessing to His people for all time.
Go in peace, my son."

After this, Abraham and Jacob spent time together, and Rebecca loved Jacob with all her heart, far more than Esau. However, Isaac continued to favor Esau, loving him more than Jacob.

Chapter XX.

In the forty-second jubilee, during the first year of the seventh week, Abraham gathered his family together. He called Ishmael and his twelve sons, Isaac and his two sons, and the six sons of Keturah along with their children.

Abraham taught them to follow the ways of the Lord, to live with honesty and kindness, and to treat others with fairness and justice. He urged them to stay true to God's commandments, never turning away from them. He warned them to avoid all kinds of immoral and impure behavior and to make sure such actions did not take place in their families or communities.

He stressed that if any woman or girl among them committed an immoral act, she should be punished, and no one should desire her or seek her out. He also warned them not to marry the daughters of Canaan because the people of Canaan would one day be removed from the land.

He reminded them about the punishment that came upon the giants and the people of Sodom. He described how they were destroyed because of their wickedness, their sins, and the corruption they spread.

Stay away from sin and anything unclean,
    And always choose to do what is right.
Do not bring shame to our family,
    Or disgrace to your own lives.

Do not let your children suffer violence,
    Or bring a curse upon yourselves like Sodom,
    Or have your descendants punished like the people of
    Gomorrah.

My sons, I urge you to love the God of heaven,
    And follow all His commandments.

Do not be led astray by false gods or their wicked ways.
Do not make idols for yourselves,
    For they are useless and lifeless,
    Created by human hands, and trusting in them is trusting in
nothing.

Do not bow down to them or serve them,
    But worship the Most High God and honor Him always.
Seek His favor and strive to do what is right,
    So that He may be pleased with you, show you kindness,
    And send rain for your fields in the morning and evening.

May He bless the work of your hands,
    Bless your food and water,
    Bless your children and your land,
    And bless your animals and flocks.

You will be a blessing to the world,
    And all nations will look to you with honor.
They will bless your children in my name,
    So they too may receive the same blessings I have been given.

Abraham gave gifts to Ishmael and his sons, as well as to the sons of Keturah. Then, he sent them away from Isaac, giving all that he owned to Isaac.

Ishmael and his sons, along with the sons of Keturah and their families, traveled together and settled in the lands stretching from Paran to the entrance of Babylon, covering the eastern regions near the desert. Over time, they intermarried and became known as the Arabs and the Ishmaelites.

Chapter XXI.

In the sixth year of the seventh week of this jubilee, Abraham called his son Isaac and said,"My son, I have grown old, and I do not know how much time I have left. I am now 175 years old, and my life has been full. Throughout the years, I have always remembered the Lord and done my best to follow His ways with all my heart. I have lived with honesty and integrity. I have rejected idols and those who worship them. My heart and soul have been fully devoted to obeying my Creator, for He is the one true God—holy, faithful, and completely just. He does not show favoritism or accept bribes. He judges fairly and will hold accountable those who break His laws or abandon His covenant.

Now, my son, follow His commandments, obey His instructions, and live according to His laws. Stay away from anything sinful or unclean, especially idol worship. Never eat the blood of any animal, whether from cattle, birds, or any other creature. If you offer a peace sacrifice, do it properly. Pour its blood on the altar and burn its fat along with fine flour mixed with oil and a drink offering. These will create a pleasing aroma to the Lord. When offering a thanksgiving sacrifice, burn the fat from the belly, the inner organs, the kidneys, and the fat near the loins and liver. Place these parts on the fire of the altar along with the meat and the drink offering as a sweet-smelling sacrifice to the Lord.

Eat the meat on the same day or the next, but never on the third day. If any remains until then, it is no longer acceptable and must not be eaten. Anyone who eats it on the third day commits a sin. I have read these instructions in the writings of our ancestors, in the words of Enoch and Noah. Also, always sprinkle salt on your offerings. The salt of the covenant must never be missing from any sacrifice you bring before the Lord.

When choosing wood for sacrifices, only use these kinds: cypress, bay, almond, fir, pine, cedar, savin, fig, olive, myrrh, laurel, or aspalathus. Pick wood that is strong, fresh, and looks good. Do not use wood that is cracked, dark, or damaged. Never use old wood, as it has lost its fragrance and will not create a pleasing aroma before the Lord. Apart from the types I've mentioned, do not use any other kind, as they do not produce a sweet scent.

Follow these instructions carefully, my son, so that you will do what is right in all things. Keep yourself clean at all times. Wash with water before approaching the altar. Wash your hands and feet before and after offering a sacrifice. Make sure no blood remains on your body or clothes. Be careful when handling blood—always cover it with dust. Never eat blood, for it holds the life of the creature. Do not consume any blood at all.

Never accept money or bribes in exchange for a person's life, so that innocent blood is not shed without justice. When blood is spilled, the land becomes polluted, and it can only be cleansed by the blood of the one responsible. Do not take payment to excuse the taking of a life. A life must be paid for with life so that you remain right before the Lord, the Most High God, who protects those who do good. May He shield you from evil and save you from all harm.

My son, I have seen that people's actions are full of sin and wickedness. Their ways are unclean, filled with evil and corruption. There is no righteousness among them. Be careful not to follow their ways or imitate their behavior. Do not commit sins that lead to death before the Most High God. If you do, He will turn away from you, allow you to fall into your own wrongdoing, and remove you and your descendants from the land. Your name and family line will be erased from the earth.

Stay far from their sinful ways and unclean actions. Follow the laws of the Most High God, obey His will, and live righteously in all things. If you do this, He will bless your work and raise a righteous family from you for generations to come. Your name and mine will never be forgotten under heaven and will last forever.

Go in peace, my son. May the Most High God—my God and your God—give you strength to do His will. May He bless your descendants and all future generations with righteousness so that you and your family will be a blessing to the whole earth."

With that, Abraham departed, his heart full of joy.

Chapter XXII.

In the second year of the first week of the forty-fourth jubilee, the same year Abraham passed away, Isaac and Ishmael traveled from the Well of the Oath to celebrate the Feast of Weeks—the festival of the first fruits of the harvest—with their father, Abraham. Seeing both of his sons together made Abraham very happy. Isaac, who owned a lot of land in Beersheba, often visited his property before returning to his father. Around that time, Ishmael also came to visit, and the two brothers reunited.

Isaac offered a burnt sacrifice on the altar that Abraham had built in Hebron. He also presented a thank offering and shared a joyful feast with his brother Ishmael. Rebecca baked fresh cakes from the new grain and gave them to Jacob to bring to Abraham so he could eat and bless the Creator before his passing. Isaac also sent a generous thank offering with Jacob for Abraham to enjoy. Abraham ate and drank, then praised the Most High God, saying:

"Blessed is the Creator of heaven and earth,
> Who made all the good things of this world,
> And gave them to people,
> So they may eat, drink, and give thanks to their Creator."

Then he continued, "I thank You, my God, for letting me see this day. I am now 175 years old, and my life has been full. I have lived in peace, and no enemy has been able to harm me in all that You have given to me and my children. My God, may Your kindness and peace be with me and my descendants. May they be a chosen people, set apart as Yours among all the nations for generations to come."

Then Abraham called Jacob and said, "My son Jacob, may the God who created everything bless you and give you the strength to live righteously and follow His ways. May He choose you and your descendants to be His people forever. Come closer, my son, and give me a kiss."

Jacob stepped forward and kissed him, and Abraham said:

"Blessed be my son Jacob,
> And blessed be all the children of the Most High God
> forever.
May God give you righteous descendants,
> And may He choose some of your children to be holy among
> the nations.
May other nations serve you,
> And may all people respect your family.
Be strong in the presence of others,
> And lead the descendants of Seth.
Through you, righteousness will continue,
> And you will become a holy nation."

"May the Most High God bless you with all the blessings He gave
    to me, to Noah, and to Adam.
May these blessings remain with your descendants forever.
May He cleanse you from all sin and impurity,
    Forgiving any wrongs you have done without knowing.
May He strengthen and bless you.
May you inherit the whole earth,
    And may He renew His covenant with you,
    Making you His chosen people forever.
May He always be your God,
    And the God of your descendants,
In truth and righteousness, for all time."

"Remember my words, my son Jacob,
    And always follow the commandments of your father,
    Abraham.
Do not mix with the other nations.
Do not eat with them or follow their customs.
Do not form close friendships with them,
    For their ways are corrupt and sinful.
They worship the dead and follow evil spirits.
They even eat meals near graves,
    And their actions are meaningless.
They do not understand,
    And they say to a piece of wood, 'You are my god,'
    And to a stone, 'You are my lord and savior.'
They are blind to the truth of their own actions."

"My son Jacob, may the Most High God guide and bless you.
May He keep you away from their wickedness and sinful ways.

108

Do not marry any of the daughters of Canaan,
>For their descendants are doomed to destruction.
Because of Ham's sin, the line of Canaan will be completely
>wiped out.
None of them will be saved on the Day of Judgment.
Idol worshipers and those who live in impurity
>Will have no place in the land of the living.
They will be forgotten on earth
>And sent to Sheol, the place of judgment,
>Just like the people of Sodom, who were completely
>destroyed."

"Do not be afraid, my son Jacob.
Be strong, my child, a descendant of Abraham.
May the Most High God protect you from harm
And rescue you from evil.
This house I have built carries my name on the earth,
>And it belongs to you and your descendants forever.
It will be known as the house of Abraham.
You will honor my name before God forever,
>And your descendants will carry my name
Through all generations of the earth."

After Abraham finished blessing Jacob and giving him instructions, they lay together on one bed. Jacob rested in Abraham's arms, and Abraham kissed him seven times, filled with love and joy for his grandson. With all his heart, he blessed Jacob, saying:

"May the Most High God, the Creator of all things,
>Who brought me out of Ur of the Chaldees to inherit this
>land forever,

Bless my holy descendants. Blessed be the Most High forever."

Then he said to Jacob, "My dear son, who brings me great joy, may God's kindness and grace always be with you and your children. May He never leave you or turn away from you. May His eyes always watch over you and your family. May He protect and bless you, choosing you as His own people. May He give you every lasting blessing, renewing His promise with you and your descendants for generations to come, according to His perfect plan."

Chapter XXIII.

Jacob rested in Abraham's arms, unaware that his grandfather had passed away. Abraham gently placed two of Jacob's fingers over his eyes, blessed the God of all gods, covered his own face, stretched out his feet, and peacefully passed away, joining his ancestors.

When Jacob woke up, he noticed that Abraham's body was cold, like ice. Alarmed, he cried out, "Grandfather, Grandfather!" But there was no response. Realizing that Abraham had died, Jacob quickly ran to his mother, Rebecca, and told her what had happened. Rebecca then went to Isaac in the night and informed him. Together with Jacob, who carried a lamp, they entered the room and found Abraham lying still, lifeless.

Isaac fell over his father's body, weeping with deep sorrow, and kissed him. Soon, the entire household was filled with the sound of mourning. Ishmael, Abraham's son, also arrived and wept for his father. Everyone in the house grieved together, crying loudly from the depths of their hearts.

Later, Isaac and Ishmael buried Abraham in the cave near his wife, Sarah. For forty days, the men of the household mourned him. This included Isaac, Ishmael, their sons, and the sons of Keturah, each grieving in their own places. When the mourning period ended, it was recorded that Abraham had lived a total of 175 years—three jubilees and four weeks of years. He had lived a full life and passed away peacefully, satisfied with his years.

In the past, people had lived much longer—up to nineteen jubilees—but after the Flood, lifespans began to shorten. No one lived that long anymore, as people aged faster, suffered more hardships, and faced the increasing wickedness of the world. Abraham was different, living a life pleasing to God, blameless in his actions. Yet, even he did not reach four full jubilees because of the growing evil on earth, which shortened his days.

From that time forward, human lifespans would continue to decrease. By the time of the final judgment, people would no longer live for even two full jubilees. As they aged, their knowledge would fade, and their understanding would weaken. A man who lived a jubilee and a half would be considered old, yet most of his life would be filled with hardship, sorrow, and suffering, with little peace.

Disaster would come one after another—wounds upon wounds, trouble upon trouble, and endless bad news. People would suffer from sickness, famine, exhaustion, war, captivity, and countless other struggles. These misfortunes would fall upon a generation filled with wickedness, a generation whose actions were sinful and corrupt.

In those days, people would complain, saying, "Our ancestors lived long and good lives—up to a thousand years. But now, we only live seventy or eighty years if we are strong, and our days are full of suffering. There is no peace in this evil generation."

Children would turn against their parents and elders, blaming them for their troubles. They would abandon the covenant that the Lord had made with their ancestors, refusing to follow His commandments or walk in His ways. They would completely turn away from God, caring only about themselves.

Everyone would chase after evil, and every word spoken would be full of lies. Their actions would be corrupt and disgusting, leading only to destruction. The earth would suffer because of their wickedness—vineyards would dry up, oil would disappear, and their unfaithfulness would bring ruin. Humanity, along with animals, livestock, birds, and sea creatures, would suffer because of the sins of mankind.

People would rise up against each other—young against old, the poor against the rich, the humble against the powerful, and beggars against rulers—all because they had abandoned the law and the covenant. They would forget the sacred commandments, the holy festivals, the Sabbaths, the jubilees, and all of the Lord's instructions.

Armed with weapons, they would fight in the hope of finding righteousness again, but they would never return to the right path until the earth was covered in blood. One person would turn against another in endless violence. Even those who survived would refuse to change. Instead, they would be filled with pride and greed, seeking only to take from others. They would claim the name of the Lord but live without truth or righteousness, corrupting even the most sacred places with their sin.

A great punishment would fall upon this generation. The Lord would allow them to fall into war, captivity, and suffering. He would bring ruthless foreign nations against them—people without mercy, who would not spare the old or the young. These invaders would be more wicked and powerful than any before them, bringing destruction to Israel and committing terrible sins against Jacob's descendants.

The land would be covered in blood, and there would be no one left to bury the dead.

In those days, people would cry out for help, calling for rescue from sinners and oppressors, but no one would come to save them.

The heads of children would turn gray with age,
> And even babies as young as three weeks old would appear as old as men of a hundred years,
> Worn down by suffering and endless hardship.

Yet in those days, some would begin to seek the Lord again.

Children would study the law,
> Search for the commandments,
> And return to the path of righteousness.

Lifespans would grow longer,
> And people would once again live for many years,
> Almost reaching a thousand years,
> Just as in the days of old.

No one would be considered old,
> And no one would feel like their life was too short.
Everyone would have the strength of youth and the joy of childhood.

Their days would be filled with peace and happiness,
> For Satan and all evil forces would be removed from the earth.
Blessings and healing would fill their lives.

In that time, the Lord would restore health to His people,
>And they would rise up in joy,
>Driving away their enemies.

The righteous would celebrate with gratitude,
>Lifting their voices in endless praise to the Lord,
>Witnessing His justice and the defeat of their enemies.

Their bodies would rest in the earth,
>But their spirits would rejoice forever,
>Knowing that the Lord is a just and merciful judge,
>Showing kindness to all generations who love Him.

Then the Lord said to Moses, "Write down these words, for they are recorded on the heavenly tablets as a message for future generations."

Chapter XXIV.

In the first year of the third week of the forty-fourth jubilee, after Abraham passed away, the Lord blessed his son Isaac. Isaac left Hebron and moved to live near the Well of the Vision, where he stayed for seven years. In the first year of the fourth week, a famine struck the land, just like the one that had happened during Abraham's time.

One day, Jacob was cooking a pot of lentils when Esau came in from the fields, exhausted and starving. Esau said, "Give me some of that red stew." Jacob replied, "Sell me your birthright, and I will give you bread and stew." Esau thought, "I'm so hungry I could die—what good is my birthright to me?" So he agreed, swearing an oath and giving his birthright to Jacob. Jacob then gave him bread and stew, which Esau ate until he was full. Esau did not care about his birthright, and because of this, he was called Edom, named after the red stew for which he had traded his inheritance. From that moment on, Jacob became the rightful firstborn, and Esau lost his position.

When the famine ended, in the second year of that week, Isaac planned to go to Egypt but instead went to Gerar, where Abimelech, the king of the Philistines, ruled. The Lord appeared to Isaac and said, "Do not go to Egypt. Stay in the land I show you and live here. I will be with you and bless you. I will give this land to you and your descendants and fulfill the promise I made to Abraham, your father. I will make your family as numerous as the stars in the sky and give them all this land. Through your descendants, all nations will be blessed because Abraham listened to My voice, obeyed My commandments, and followed My laws. Now, you must do the same and remain here."

Isaac stayed in Gerar for 21 years. While he was there, Abimelech warned his people, "Anyone who harms Isaac or takes anything that belongs to him will be put to death." Isaac became successful among the Philistines, gaining many possessions, including oxen, sheep, camels, donkeys, and many servants. He planted crops in the Philistine land and harvested a hundred times more than expected, becoming very wealthy. This made the Philistines jealous.

Out of envy, the Philistines filled the wells that Abraham's servants had dug with dirt. Then Abimelech said to Isaac, "Leave us because you have become too powerful for us." So in the first year of the seventh week, Isaac moved to the valleys of Gerar. There, his servants reopened the wells that Abraham had dug, which the Philistines had filled after his death. Isaac named the wells just as his father had.

Isaac's servants dug a new well in the valley and found fresh water. But the shepherds of Gerar argued with them, saying, "This water belongs to us." So Isaac named the well "Dispute" because of their unfair claim. His servants dug another well, but the locals fought over it too, so he named it "Opposition."

Isaac then moved on and dug another well, and this time, no one fought over it. He named it "Room" and said, "The Lord has made space for us, and we have prospered in this land."

In the first year of the first week of the forty-fourth jubilee, Isaac moved to the Well of the Oath. That night, on the new moon of the first month, the Lord appeared to him and said, "I am the God of Abraham, your father. Do not be afraid, for I am with you. I will bless you and make your descendants as numerous as the sand of the earth because of My servant Abraham." Isaac built an altar at the same place where his father had built one. He called on the name of the Lord and offered a sacrifice. His servants then dug another well and found fresh water.

Later, Isaac's servants dug another well but found no water. They told Isaac, and he said, "I have sworn peace with the Philistines, and they know about this well." He named the place "Well of the Oath" because of the peace agreement he had made with Abimelech, Ahuzzath, and Phicol, the commander of their army. Isaac understood that he had no choice but to make peace with them.

On that day, Isaac cursed the Philistines, saying:

"The Philistines will be cursed until the day of judgment when they are scattered among the nations. They will be a disgrace and a curse, hated and punished by sinful nations and the Kittim. Even if they escape the enemy's sword and the Kittim, they will still face judgment from the righteous nation. They will remain enemies of my descendants forever. When the day of judgment arrives, their entire bloodline will be wiped out, and no one from the Caphtorim will be left on earth.

If they try to rise to power, they will be brought down. If they become strong, they will be uprooted. If they hide among other nations, they will be found and removed. If they descend into Sheol, they will suffer greatly and never find peace. If they are taken as captives, those who pursue them will destroy them before they can escape. No descendants will remain, and their name will be erased from history. They will be cursed forever."

This decree is written on the heavenly tablets and will be fulfilled on the day of judgment, when the Philistines will be completely wiped out from the earth.

Chapter XXV.

In the second year of that week, during this jubilee, Rebecca called her son Jacob and said, "My son, do not marry one of the daughters of Canaan like your brother Esau. He has married two women from this land, and they have brought me nothing but sorrow. Their ways are immoral and sinful, and there is no goodness in them. Everything they do is filled with wickedness, and they bring only grief and pain.

I love you deeply, my son, and I bless you every moment, day and night. Please listen to me and follow my advice. Do not choose a wife from the women of this land. Instead, find a wife from my father's family, from among our own people. If you marry within our family, the Most High God will bless you, and your children will be righteous and holy."

Jacob answered his mother, "Mother, I am still young, only nine weeks old, and I have no knowledge or experience with women. I have not made any promises to anyone, and I do not plan to marry a daughter of Canaan.

I remember what our father Abraham told me. He warned me not to marry a woman from Canaan but to choose a wife from our own family and people. I know that your brother Laban has daughters, and I would like to marry one of them.

I have kept myself pure for this reason, staying away from sin and corruption. Father Abraham gave me clear instructions to avoid lust and wrongdoing, and I have followed his guidance.

Even though my brother Esau has pressured me many times to marry one of his wives' sisters, I have refused to follow his example. I promise you, mother, I will never marry a woman from Canaan, and I will not act wickedly as my brother has done.

Do not worry, mother. I will honor your wishes and live a righteous life, never straying from the right path."

Hearing this, Rebecca lifted her eyes to heaven, stretched out her hands, and prayed, thanking and praising the Most High God, the Creator of heaven and earth. She said,

"Blessed be the Lord God, and may His holy name be praised forever. He has given me Jacob, a pure son, a righteous seed. He belongs to You, Lord, and his descendants will be Yours for all generations to come.

Bless him, O Lord, and let my words carry the blessing of righteousness as I speak over him."

Then, filled with the spirit of righteousness, she placed her hands on Jacob's head and said,

"Blessed are You, Lord of righteousness and God of all ages. May You bless Jacob above all the people of the earth.

Guide him, my son, on the path of righteousness, and may his descendants also walk in truth and goodness.

May his children multiply like the months of the year and become as numerous as the stars in the sky, outnumbering the grains of sand by the sea.

Give them this good land, as You promised to Abraham and his descendants, and may they possess it forever.

May I live to see your children, my son, and may all your descendants be holy and blessed.

Just as you have brought me joy and comfort, may you be blessed by the womb that bore you, by the love of my heart, by the milk that nourished you, and by the words of my mouth that praise you always.

May you grow and spread across the earth, and may your descendants rejoice forever in heaven and on earth.

May your children be perfect, full of joy, and find peace on the great day of peace.

May your name and your family line last forever. May the Most High God always be your God, and may the God of righteousness be with your descendants. Through them, may His holy place be established for all generations.

Blessed are those who bless you, and cursed are those who falsely curse you."

Rebecca kissed Jacob and said, "May the Lord of all creation love you as much as my heart loves you. May my joy and blessings always remain with you." Then, she finished her blessings.

Chapter XXVI.

In the seventh year of that week, Isaac called his older son, Esau, and said, "My son, I am old, and my eyesight is failing. I do not know how much longer I will live. Take your bow and arrows, go to the fields, and hunt some wild game for me. Prepare my favorite meal and bring it to me so that I may eat and bless you before I die."

Rebecca overheard Isaac speaking to Esau. After Esau left early to hunt, she called Jacob and said, "I heard your father tell Esau, 'Go hunt for me and prepare my favorite meal so I can eat and bless you before the Lord before I die.' Now listen carefully, my son. Go to the flock and bring me two young goats. I will prepare them just the way your father likes. Then you will take the food to him, and he will eat and bless you before he dies. This way, you will receive the blessing instead."

Jacob hesitated and said, "But, Mother, what if my father recognizes my voice or touches me? I have smooth skin, and Esau is hairy. If he touches me, he will know I am tricking him, and instead of a blessing, I will bring a curse upon myself."

Rebecca replied, "Let any curse fall on me, my son. Just do what I tell you."

So Jacob obeyed and brought the two young goats. Rebecca prepared the food exactly how Isaac liked it. Then she took Esau's best clothes, which she had in the house, and dressed Jacob in them. She covered his hands and the smooth part of his neck with goat skins so that he would feel hairy like Esau. Then she handed Jacob the food and bread she had made.

Jacob went to his father and said, "Father, I have done as you asked. Please sit up and eat the meal I prepared so that you may bless me."

Isaac asked, "How did you find the game so quickly, my son?"

Jacob answered, "Because the Lord your God helped me succeed."

Isaac said, "Come closer so I can touch you, my son, and see if you really are Esau."

Jacob moved closer, and Isaac touched him. He said, "The voice is Jacob's, but the hands feel like Esau's." Isaac did not recognize him because his hands felt hairy like Esau's. It was by God's will that Isaac could not tell the difference, so he continued to bless him.

"Are you really my son Esau?" Isaac asked.

Jacob answered, "Yes, I am."

Isaac said, "Bring me the food so I may eat and bless you."

Jacob brought him the meal and wine, and Isaac ate and drank. Then Isaac said, "Come closer, my son, and kiss me."

When Jacob leaned in to kiss him, Isaac smelled the scent of Esau's clothes and blessed him, saying:

"My son smells like a field that the Lord has blessed.
    May God give you rain from the sky,
    Rich soil,
    And plenty of grain and wine.

May nations serve you,
    And people show you honor.
Be a ruler over your brothers,
    And may your mother's sons respect you.

Anyone who curses you will be cursed,
    And anyone who blesses you will be blessed."

Just as Jacob left, Esau returned from his hunt. He prepared the meal and brought it to his father, saying, "Father, sit up and eat the venison I've brought so you can bless me."

Isaac asked, "Who are you?"

Esau replied, "I am your firstborn son, Esau."

Isaac trembled and said, "Then who was it that hunted game and brought it to me? I already ate it, and I have blessed him—and the blessing will stand!"

When Esau heard this, he let out a loud and bitter cry. "Father, bless me too!" he pleaded.

Isaac said, "Your brother came with deceit and took your blessing."

Esau cried out, "No wonder his name is Jacob! He has cheated me twice—first, he took my birthright, and now he has taken my blessing! Haven't you saved any blessing for me?"

Isaac replied, "I have made him lord over you and given him all his brothers as servants, along with plenty of grain and wine. What more can I give you, my son?"

Esau begged, "Father, do you only have one blessing? Bless me too!" And he wept loudly.

Isaac said,

"You will live far from the fertile land
    And without the blessing of rain from above.
You will survive by the sword
    And serve your brother.

But when you can no longer bear it,
    You will break free from his control.
However, your choices will lead to great sin,
    And your descendants will be lost."

Esau was filled with hatred toward Jacob because of the blessing his father had given him. He said to himself, "Soon my father will die, and then I will kill my brother Jacob."

Chapter XXVII.

Rebecca had a dream warning her about Esau's plan to take revenge and kill Jacob. She immediately called Jacob and said, "Your brother Esau is planning to harm you. Listen to me, my son, and do what I say. Leave right away and go to my brother Laban in Haran. Stay there for a while until Esau calms down and forgets what happened. Then I'll send for you to return."

Jacob replied, "I'm not afraid of him. If he comes after me, I'll defend myself and fight back."

But Rebecca said, "I don't want to lose both of you in one day."

Jacob answered, "You know that Father is old and nearly blind. If I leave without his blessing, it will upset him, and he might curse me instead of bless me. I won't go unless he sends me himself."

Rebecca reassured him, "I will talk to him, and he will bless you before you leave."

She went to Isaac and said, "I am deeply troubled by Esau's wives. If Jacob marries a woman like them, I don't want to live anymore. The women of this land are full of wickedness."

Isaac then called Jacob, blessed him, and said, "Do not marry a woman from Canaan. Go to Mesopotamia, to your mother's family, and find a wife from the daughters of your uncle Laban. May God Almighty bless you, give you many children, and make you into a great nation. May He grant you the blessings promised to Abraham, and may you inherit this land, where you now live as a foreigner—the land God gave to Abraham. Go in peace, my son."

Isaac sent Jacob on his way, and Jacob traveled to Mesopotamia, to the house of Laban, the son of Bethuel, Rebecca's brother.

After Jacob left, Rebecca was filled with sorrow and wept for her son.

Isaac comforted her, saying, "Do not cry for Jacob, my love. He left in peace, and he will return in peace. The Most High God will watch over him, protect him, and guide him. He will succeed in all he does, and when he comes back, we will rejoice with him again. Be at ease, for Jacob is righteous and follows the path of truth. He will not perish. Wipe your tears."

Isaac's words reassured Rebecca, and they both blessed Jacob.

Jacob left the well of Beer-sheba and began his journey toward Haran in the first year of the second week of the 44th jubilee. He arrived at a place called Luz, later known as Bethel, on the new moon of the first month. As the sun was setting, he turned off the road and decided to spend the night there.

He took a stone from the area, placed it under his head as a pillow, and lay down to sleep.

That night, Jacob had a dream. He saw a ladder reaching from the ground up to heaven, with angels going up and down on it. At the top of the ladder stood the Lord, who spoke to him:

"I am the Lord, the God of your grandfather Abraham and the God of your father Isaac. The land where you are lying, I will give to you and your descendants. Your family will be as countless as the dust of the earth, spreading in all directions—west, east, north, and south. Through you and your descendants, all the families of the earth will be blessed.

I am with you and will protect you wherever you go. I will bring you back to this land, and I will not leave you until I have done everything I have promised."

When Jacob woke up, he said, "Surely, the Lord is in this place, and I didn't even realize it." Filled with awe, he added, "This place is amazing. It is truly the house of God and the gateway to heaven."

Early the next morning, Jacob took the stone he had used as a pillow, stood it upright as a pillar, and poured oil over it. He named the place Bethel, though it was originally called Luz.

Then Jacob made a vow, saying, "If God will be with me, protect me on this journey, provide food and clothing, and bring me back safely to my father's house, then the Lord will be my God. This stone I have set up will be the house of God, and I will give a tenth of everything You bless me with."

Chapter XXV III.

Jacob continued his journey and arrived in the land of the east, at the home of Laban, his mother Rebecca's brother. He stayed there and worked for seven years so he could marry Rachel, Laban's daughter. After completing the seven years, Jacob said to Laban, "I have worked as promised. Now give me Rachel as my wife."

Laban agreed and prepared a wedding feast, but instead of giving Rachel to Jacob, he gave him his older daughter, Leah. He also gave Leah a servant named Zilpah to be her maid. Jacob, unaware of the trick, spent the night with Leah, believing she was Rachel. When he realized the truth, he was furious and said to Laban, "Why have you deceived me? I worked for Rachel, not Leah. This is wrong! Take Leah back and let me go."

Jacob loved Rachel more than Leah. Leah had soft eyes, but Rachel was very beautiful and had a graceful figure. Laban told Jacob, "It is not our custom to marry off the younger daughter before the older one.

This rule is also written in the heavenly records—it is a sin to break it." He warned Jacob never to go against this law in the future.

Laban then said, "Finish the wedding celebrations for Leah, and after one week, I will also give you Rachel. But in return, you must work for me another seven years."

Jacob agreed, and after the seven-day celebration for Leah, Laban gave Rachel to him as well. He also gave Rachel a servant named Bilhah to be her maid. Jacob then worked another seven years to fulfill his promise for Rachel.

The Lord saw that Leah was unloved, so He blessed her with children. In the first year of the third week, Leah gave birth to a son and named him Reuben. But Rachel remained childless because the Lord had not yet given her children, as Jacob loved her more than Leah.

Leah became pregnant again and had a second son, Simeon. Later, she had a third son and named him Levi. Then she gave birth to a fourth son and called him Judah.

Meanwhile, Rachel became jealous of Leah for having children and said to Jacob, "Give me children, or I will die!"

Jacob answered, "I am not God. He alone decides who can have children."

Seeing that Leah had four sons—Reuben, Simeon, Levi, and Judah—Rachel said to Jacob, "Take my servant Bilhah as a wife. She will have children for me."

Rachel gave Bilhah to Jacob, and Bilhah became pregnant and had a son. Rachel named him Dan. Bilhah became pregnant again and had another son, whom Rachel called Naphtali.

Leah noticed that she had stopped having children, so she gave her servant Zilpah to Jacob as a wife. Zilpah gave birth to a son, and Leah named him Gad. Then Zilpah had another son, and Leah named him Asher.

Later, Leah became pregnant again and gave birth to a son named Issachar. She had another son, whom she called Zebulun, and then a daughter, Dinah.

Finally, the Lord answered Rachel's prayers and allowed her to have a child. She became pregnant and gave birth to a son, naming him Joseph.

After Joseph was born, Jacob said to Laban, "Let me take my wives and children and return to my father's house. I have served you faithfully, and now I wish to go home."

Laban replied, "Stay with me a little longer. Name your wages, and I will pay you. Continue caring for my flocks."

They agreed that Jacob's payment would be all the black, spotted, and speckled lambs and goats born among Laban's flocks. Over time, more and more animals were born with these markings, increasing Jacob's share.

Jacob's wealth grew, and he gained many sheep, oxen, camels, donkeys, servants, and maids. However, Laban and his sons became jealous of Jacob's success. Laban started taking back some of the sheep and treated Jacob unfairly.

Chapter XXIX.

After Rachel gave birth to Joseph, Laban left to shear his sheep, which were three days away from his home. Seeing this as the perfect time to leave, Jacob called Leah and Rachel and spoke to them kindly, asking them to go with him back to the land of Canaan. He told them about a dream in which God instructed him to return to his father's house.

Leah and Rachel agreed, saying, "We will go wherever you go."

Jacob then praised the God of his father, Isaac, and his grandfather, Abraham. Early in the morning, he gathered his wives, children, and all his belongings and crossed the river. Without telling Laban, they set out toward Gilead.

In the seventh year of the fourth week, on the twenty-first day of the first month, Jacob began his journey to Gilead. When Laban realized Jacob had left, he chased after him and caught up in the mountains of Gilead on the thirteenth day of the third month. However, God appeared to Laban in a dream that night and warned him not to harm Jacob.

On the fifteenth day, Jacob prepared a feast for Laban and his men. During the gathering, Jacob and Laban made an agreement, promising never to cross the mountains of Gilead with the intent to harm one another. To mark their promise, Jacob built a mound of stones as a witness to their covenant. The place was named "The Heap of Witness."

Before this, the region of Gilead had been home to the Rephaim, a race of giants who were between seven and ten cubits tall. Their land stretched from the territory of the Ammonites to Mount Hermon, including cities like Karnaim, Ashtaroth, Edrei, Misur, and Beon. Because of their wickedness, God wiped them out, and the Amorites later took over their land. No people since have sinned to the same extent as the Rephaim, and their time on earth was cut short.

After making peace, Jacob sent Laban on his way, and Laban returned to Mesopotamia. Jacob continued his journey and crossed the Jabbok River on the eleventh day of the ninth month. That same day, his brother Esau came to meet him. The two brothers reconciled and made peace. Esau then traveled back to the land of Seir, while Jacob remained living in tents.

In the first year of the fifth week, Jacob crossed the Jordan River and settled on the other side. He grazed his flocks from the Sea of the Heap to Bethshan, Dothan, and the Akrabbim forest. From his wealth, Jacob sent gifts to his father Isaac, including clothes, food, meat, drink, milk, butter, cheese, and dates from the valley.

He also sent gifts to his mother Rebecca four times a year—after plowing, during harvest, after autumn, and in the spring. He delivered them to the tower of Abraham, where Rebecca lived.

Meanwhile, Isaac had moved from the Well of the Oath to the tower of Abraham in the mountains of Hebron, living separately from Esau. During Jacob's time in Mesopotamia, Esau married Mahalath, Ishmael's daughter. Later, Esau moved to Mount Seir with his flocks and wives, leaving Isaac behind at the Well of the Oath. Isaac then relocated to the tower of his father, Abraham.

Jacob continued to provide for his father and mother, sending them whatever they needed. In return, Isaac and Rebecca blessed him with all their hearts and souls.

Chapter XXX.

In the first year of the sixth week, during the fourth month, Jacob settled in Salem, east of Shechem, arriving safely. While there, Shechem, the son of Hamor the Hivite, the ruler of the land, took Jacob's young daughter Dinah into his house and violated her. She was only twelve years old. Afterward, Shechem wanted to marry Dinah, so he approached his father and then Jacob and his sons with a proposal.

Jacob and his sons were furious about what had been done to Dinah. Although they pretended to respond peacefully, they secretly planned revenge.

Simeon and Levi launched a surprise attack on Shechem, killing all
the men in the city to avenge their sister. They left no one alive, making
it clear that no daughter of Israel should ever be treated this way again.
It was declared in heaven that anyone who committed such a crime
deserved to die, for their actions brought disgrace to Israel.

The Lord allowed Jacob's sons to carry out this judgment to
prevent others from committing the same sin in the future. It was also
commanded that if any Israelite man gave his daughter or sister to a
foreigner, he would be put to death by stoning. Likewise, any Israelite
woman who dishonored her family by marrying outside of Israel would
be burned and removed from the nation.

Israel was called to remain pure before the Lord, avoiding all
impurity and unfaithfulness. Anyone who corrupted the nation would
face death, as recorded in the heavenly laws. No forgiveness or
atonement would be granted to those who violated these laws. Anyone
who allowed impurity to spread or gave their children to foreign
customs was committing a serious sin and would be cut off from the
nation.

Moses was instructed to warn the people of Israel never to marry
Gentiles or allow their daughters to do so, for it was considered a
terrible offense before the Lord. The story of Shechem was recorded
as a warning, showing the judgment carried out by Jacob's sons, who
declared, "We will never give our sister to an uncircumcised man, for
it would be a disgrace to us."

Marrying outside of Israel was seen as sinful and shameful. Such
actions would bring plagues and curses upon the nation. Anyone who
ignored this impurity or allowed it to happen would also face judgment
and punishment. The entire community would suffer because of it, and

the Lord would reject any offerings, sacrifices, or incense from those who committed these sins.

This is why Moses was commanded to teach Israel about the Shechemites and how they were judged by Jacob's sons. Their actions were seen as righteous, and they were written in the heavenly records as a blessing. Levi, in particular, was honored for his dedication to justice. Because of this, he and his descendants were chosen to serve as priests for the Lord forever. His name was recorded in the heavenly tablets as a righteous man and a faithful servant of God.

These events were recorded to remind Israel of their covenant and the importance of following God's laws. If they obeyed, they would be counted as friends of the Lord. But if they sinned and broke the covenant, they would be seen as enemies, erased from the book of life, and placed among those destined for destruction.

On the day Jacob's sons destroyed Shechem, their actions were recorded in heaven as righteous judgment. They rescued Dinah from Shechem's house and took everything from the city—the livestock, goods, and wealth—including sheep, oxen, and donkeys—bringing it all back to Jacob.

Although Jacob was upset with his sons for attacking the city, fearing that the neighboring Canaanites and Perizzites would seek revenge, the fear of the Lord fell upon those nearby. No one dared to attack Jacob's family because terror had spread among the surrounding cities.

Chapter XXXI.

At the beginning of the month, Jacob gathered his family and said, "Purify yourselves and change into clean clothes. We are going to Bethel, the place where I made a vow to God when I fled from my brother Esau. God has been with me, protected me, and brought me safely back to this land. Now, get rid of any foreign gods among you."

His family handed over their idols, including the ones Rachel had taken from her father Laban, as well as the jewelry they wore in their ears. Jacob collected everything, destroyed them, and buried the remains under an oak tree in Shechem.

Jacob then traveled to Bethel at the beginning of the seventh month. He built an altar at the place where he had once slept and set up a pillar in honor of the Lord. He sent a message to his father, Isaac, inviting him and his mother, Rebekah, to join him for the sacrifice. Isaac replied, "Let my son Jacob come to me so I may see him before I die."

Jacob brought his sons, Levi and Judah, to visit Isaac. When Rebekah heard that Jacob had arrived, she left the tower and ran to meet him. Her heart filled with joy when she heard, "Jacob, your son, has returned." She embraced him tightly, kissed him, and wept with happiness. Then she saw Levi and Judah and asked, "Are these your sons, my child?" She hugged and kissed them, blessing them, and said, "Through you, the descendants of Abraham will become a great people, bringing blessings to the world."

Jacob entered Isaac's room, where his father lay resting. He took Isaac's hand, leaned down, and kissed him. Isaac held Jacob close and wept. Though his eyesight was failing, his spirit lifted, and he asked, "Are these your sons? They look just like you." Jacob confirmed that they were.

Isaac kissed both boys, and suddenly, the spirit of prophecy filled him. He took Levi's right hand and Judah's left, then began to bless Levi first.

"May the eternal God bless you and your children. May you be chosen and set apart to serve in His holy place, just like the angels in heaven. Your family will be honored and respected, always serving in God's presence. Your descendants will lead and guide the tribes of Israel, teaching them His laws and commandments. His blessings will be spoken through you, and you will bring goodness to all His people."

Isaac continued, "Your mother named you Levi, and your name is fitting because you will always be connected to the Lord. You will share in the inheritance with Jacob's sons, and your family will receive blessings from God's table. May you never be in need and always have more than enough. Anyone who stands against you will fail, and those who curse you will be cursed, but those who bless you will be blessed."

Then Isaac turned to Judah and said, "May the Lord give you the strength to defeat your enemies. You will be a leader among your brothers, and from your family will come a ruler who will govern the descendants of Jacob. Your name will be known everywhere, and nations will respect your family. Through you, Jacob's people will find help, and Israel will be saved. When you rule with fairness, peace will spread to all of God's people. Those who bless you will be blessed, and those who stand against you will fade away."

Isaac kissed Judah, hugged him, and felt great joy at seeing Jacob's sons. He blessed them once more and then rested at Jacob's feet. That evening, Isaac and Jacob ate together, celebrating with happiness. Jacob placed Levi and Judah on either side of Isaac as they slept, and Isaac felt it was a righteous act.

That night, Jacob shared stories with Isaac about how the Lord had been with him, protected him, and blessed him. Isaac praised the God of Abraham for showing mercy to Jacob and his family.

The next morning, Jacob told Isaac about the vow he had made at Bethel, describing the vision he had seen and the altar he had built. Isaac said, "I am too old to travel, my son. I am 165 years old and no longer strong enough to make the journey. Take your mother with you and fulfill the vow you made to the Lord without delay. Be faithful to it, for you are responsible for keeping your promise. May the Creator of all things accept your offering and be pleased."

Isaac instructed Rebekah to go with Jacob, and she traveled with him, bringing her servant Deborah. As they made their way to Bethel, Jacob reflected on Isaac's blessings over Levi and Judah, and his heart was filled with gratitude. He praised the God of his fathers, Abraham and Isaac, saying, "Now I know that my future is secure, and that my sons will also be blessed before the Lord forever."

This moment was recorded in the heavenly books as an eternal testimony of Isaac's blessings over Levi and Judah.

Chapter XXXII.

That night, Jacob stayed in Bethel, and Levi had a dream where God chose him and his descendants to serve as priests for the Most High forever. When he woke up, Levi praised and thanked God for this great honor.

The next morning, on the fourteenth day of the month, Jacob dedicated a tenth of everything he owned—his servants, livestock, gold, silver, and all his possessions—as an offering to God. Around this time, Rachel became pregnant with her son, Benjamin. Jacob counted his sons and decided that Levi's share would be set apart for the Lord. He dressed Levi in priestly robes and gave him the responsibility of serving as a priest.

On the fifteenth day of the month, Jacob made a sacrifice at the altar, offering fourteen oxen, twenty-eight rams, forty-nine sheep, seven lambs, and twenty-one goats, along with grain and drink offerings. This was to fulfill a promise he had made to God. After the fire consumed the sacrifices, Jacob burned incense on the altar as an offering of thanksgiving.

For the next seven days, he continued making offerings, sacrificing two oxen, four rams, four sheep, four goats, and two lambs each day. Jacob, his sons, and his household celebrated with food and drink, thanking God for His kindness and protection.

Jacob also offered a tenth of all his clean animals as a burnt offering to the Lord. However, the unclean animals were not included in Levi's portion. Levi served as a priest at Bethel, leading the offerings before Jacob and his brothers. Jacob renewed his promise to God, dedicating another tithe and setting it apart for the Lord. This became a lasting tradition: every year, the second tithe was to be eaten in the place God chose, with none of it left over. Any tithe not eaten by the end of the year would be considered unclean and burned.

Jacob then planned to build a sacred place at Bethel, enclosing it with a wall to create a permanent sanctuary for himself and his descendants. That night, the Lord appeared to him, blessed him, and said, "Your name will no longer be Jacob but Israel." God promised that Jacob's family would grow into many nations and that kings would come from his descendants. He also reaffirmed that the land under heaven would belong to Jacob's children, who would rule over many nations.

After God finished speaking, Jacob watched as He ascended into heaven. Later that night, Jacob had another dream where an angel came down from heaven carrying seven tablets. The angel handed them to

Jacob, and as he read them, he understood everything that would happen to him and his descendants in the future.

The angel told Jacob not to build a permanent sanctuary at Bethel but to return to his father's house until Isaac passed away. The angel also revealed that Jacob would die peacefully in Egypt and be buried with honor alongside Abraham and Isaac.

The angel reassured Jacob not to be afraid, promising that everything he had seen and read would come true. When Jacob worried about remembering all the details, the angel told him that he would recall them when the time was right.

After the vision ended, Jacob woke up, remembering everything he had seen. He wrote it all down and, the next day, made another offering as he had before. He named the day "Addition" because it was added to the yearly feast days, and it became a part of Israel's yearly celebrations.

On the night of the twenty-third of the month, Deborah, Rebekah's nurse, passed away. Jacob buried her near a river by the city, under an oak tree, naming the place "The River of Deborah" and the tree "The Oak of Deborah's Mourning."

Rebekah returned home with Isaac, and Jacob sent gifts of rams, sheep, and goats for them to prepare a meal. Later, Jacob traveled to the land of Kabratan to be near his mother and stayed there for a while.

During this time, Rachel gave birth to a son. Her labor was very painful, and she named him "Son of My Sorrow." However, Jacob changed his name to Benjamin. Sadly, Rachel died during childbirth, and Jacob buried her in Ephrath, which is also called Bethlehem. He set up a pillar on her grave as a marker, and it remained there along the road.

Chapter XXXIII.

Jacob settled with his wife Leah south of Magdaladra'ef on the first day of the tenth month. Around this time, Reuben, Jacob's oldest son, saw Bilhah, Rachel's servant and his father's concubine, bathing in a private place. He was attracted to her and waited until nighttime to act on his desires.

Late at night, Reuben secretly entered Bilhah's tent while she was sleeping alone. He lay with her, and when she woke up and realized what had happened, she was horrified. Bilhah screamed when she recognized Reuben and clung to her blanket in shame. Reuben ran away, leaving her deeply distressed. She was overcome with grief but kept what had happened to herself.

When Jacob returned to Bilhah, she told him everything, saying, "I am no longer pure for you because I have been dishonored. Reuben lay with me while I was asleep, and I didn't know until he uncovered my blanket and defiled me." Jacob was furious with Reuben for committing such a terrible sin, bringing shame upon his father. From that day forward, Jacob never went to Bilhah again.

This act was a great sin in God's eyes. It is strictly forbidden for a man to be with his father's wife or bring shame upon his father in this way. Such behavior is disgraceful and offensive to the Lord. It is written in the heavenly records that no man should ever commit this sin. The punishment for it is death by stoning, and both the man and the woman involved must be removed from the people of God to keep the nation pure.

It is also written, "Cursed is the man who lies with his father's wife, for he has shamed his father." And all the holy ones of the Lord declared, "Amen." Moses was commanded to teach this law to the children of Israel, for it carries the penalty of death. It is a serious impurity, and there is no forgiveness for it. Anyone who commits this sin must be put to death immediately and removed from the community. Such a person must not remain alive even for a single day, as their actions have polluted the people of God.

Although Reuben was not punished while Jacob was alive, it was only because the full law and its judgment had not yet been given. Now, this law is established forever for all generations. There is no atonement for this sin. Anyone who commits it must be removed from the nation and put to death on the same day. Moses was commanded to write down this law so that Israel would follow it and avoid sins that lead to destruction.

The Lord, who is a just Judge, does not show favoritism and cannot be bribed. Moses was to remind the people of this covenant so they would obey it and protect themselves from being cut off from the land. Anyone who commits this sin is considered unclean before God. There is no greater impurity on earth than this kind of wrongdoing. Israel is a holy nation, chosen by God to receive His promises—a people set apart to serve Him. Such impurity must not be found among them.

In the third year of the sixth week, Jacob and all his sons moved to the home of Abraham, near Isaac, his father, and Rebekah, his mother. Jacob's sons were: Reuben, the firstborn; Simeon, Levi, Judah, Issachar, and Zebulun, the sons of Leah; Joseph and Benjamin, the sons of Rachel; Dan and Naphtali, the sons of Bilhah; Gad and Asher, the sons of Zilpah; and Dinah, Leah's only daughter.

When they arrived, they bowed in respect before Isaac and Rebekah, and Isaac blessed Jacob and his sons. Seeing Jacob's children filled Isaac with joy. He spoke blessings over them, praising God for the family that had grown through Jacob.

Chapter XXXIV.

In the sixth year of the forty-fourth jubilee, Jacob sent his sons and their servants to graze the sheep near Shechem. While they were there, seven kings of the Amorites made a plan to attack them. They hid among the trees, waiting for the right moment to steal their livestock.

At that time, Jacob was at home with Levi, Judah, Joseph, and Isaac because Isaac was feeling sorrowful, and they didn't want to leave him alone. Benjamin, the youngest, also stayed with his father.

The kings of Taphu, Aresa, Seragan, Selo, Ga'as, Bethoron, and Ma'anisakir, who were from Canaan, learned about the Amorites' plot and sent a message to Jacob: "The Amorite kings have surrounded your sons and taken their herds."

Jacob immediately gathered Levi, Judah, Joseph, his father's servants, and his own men, making a total of six thousand warriors armed with swords. They marched to Shechem to face the Amorites. A fierce battle broke out, and Jacob's forces chased down and defeated the Amorites. They killed the kings of Aresa, Taphu, Seragan, Selo, Ga'as, and Ma'anisakir, reclaimed the stolen livestock, and subdued their enemies.

After the battle, Jacob forced the defeated kings to pay tribute by giving five types of fruit from their land. He also built two cities, Robel and Tamnatares, before returning home safely. The defeated kings remained under Jacob's rule until he and his sons later moved to Egypt.

In the seventh year of the next week, Jacob sent Joseph to check on his brothers in Shechem. However, Joseph found them in Dothan, where they had moved their flocks. When Joseph arrived, his brothers plotted against him, intending to kill him. But instead of going through with their plan, they decided to sell him to Ishmaelite merchants. These merchants took Joseph to Egypt and sold him to Potiphar, Pharaoh's chief cook and a priest in the city of 'Elew.

To cover up their actions, Joseph's brothers killed a goat, dipped his coat in its blood, and sent it to Jacob on the tenth day of the seventh month. When Jacob saw the blood-stained coat, he believed that a wild animal had killed Joseph. Overcome with sorrow, he cried, "A wild beast has devoured Joseph!" His entire household mourned with him that day, but Jacob refused to be comforted, saying, "I will grieve for my son until I go to my grave."

That same month, Bilhah, who was living in Qafratef, was so heartbroken after hearing about Joseph's death that she passed away. Soon after, Dinah, Jacob's daughter, also died. Within a single month, Jacob lost Bilhah, Dinah, and Joseph—or so he believed. Both Bilhah and Dinah were buried near Rachel's grave.

Jacob mourned for Joseph for an entire year and could not find peace. Again and again, he repeated, "I will go to my grave mourning for my son."

Later, it was commanded that the people of Israel fast on the tenth day of the seventh month—the day Jacob learned of Joseph's supposed death. This was to be a day of atonement for their sins, marked by the sacrifice of a young goat. It also served as a reminder of the deep sorrow Jacob felt for Joseph and was set as a time for spiritual cleansing.

After Joseph was gone, Jacob's sons began to marry.

- Reuben married Ada.
- Simeon first married Adlba'a, a Canaanite, but later repented and married another wife from Mesopotamia, following the example of his brothers.
- Levi married Melka, a daughter of Aram from Terah's family.

- Judah married Betasu'el, a Canaanite.
- Issachar married Hezaqa.
- Zebulun married Ni'iman.
- Dan married Egla.
- Naphtali married Rasu'u from Mesopotamia.
- Gad married Maka.
- Asher married Ijona.
- Joseph later married Asenath, an Egyptian.
- Benjamin married Ijasaka.

Chapter XXXV.

In the first year of the first week of the forty-fifth jubilee, Rebecca called Jacob to her and gave him advice about his father and brother. She told him to always respect and care for them.

Jacob replied, "I will do everything you ask, Mother. Treating them with respect will bring me blessings and favor from God. You know my heart and how I have lived my life. I have always tried to do what is right for others. How could I not honor my father and brother as you want? If I have done anything wrong, please tell me, and I will correct it so that God will have mercy on me."

Rebecca said, "My son, I have never seen you do anything wrong, only good. But I must tell you something important. My time is near, and I will die this year. I will not live beyond the age of 155. I have seen it in a dream, and I know it is true."

Jacob laughed at her words because Rebecca was still strong and healthy. She had no sign of illness, moved around easily, and had never been sick. Jacob said, "Mother, if I could live as long as you and still be as strong as you are, I would consider it a blessing. You won't die; you must be joking with me."

Rebecca then went to Isaac and said, "I have a request, my husband. Please make Esau promise that he will not harm Jacob or stay angry with him. You know how Esau has always been—he has been difficult since childhood, and there is no kindness in him. After you pass away, he plans to kill Jacob. You have seen how he has treated us, especially after Jacob left for Haran. He took your flocks and stole from us, and when we asked for what was rightfully ours, he acted like he was doing us a favor. He is still upset because you blessed Jacob, who is honest and righteous. But since Jacob came back, he has cared for us in every way. He shares what he has, respects us, and treats us with kindness."

Isaac replied, "I know everything Jacob has done. He honors us with all his heart. I used to love Esau more because he was my firstborn, but now I love Jacob more because Esau has chosen a sinful path. He does not follow what is right; he has become violent and corrupt. He has turned away from the God of Abraham and follows the ways of his wives, who have led him into impurity. Neither he nor his descendants will be saved; they will be destroyed. As for making Esau promise, even if he does, he will not keep his word. But do not worry about Jacob. He is protected by Someone far greater than Esau's strength."

Rebecca then called Esau to her. When he arrived, she said, "I have a request, my son. Will you promise to do what I ask?"

Esau answered, "I will do whatever you ask, Mother. I will not refuse you."

Rebecca said, "When I die, bury me near Sarah, your father's mother. Also, love your brother Jacob and do not harm him. If you and Jacob love each other, you will both prosper and be honored in this land. No enemy will have power over you, and you will be a blessing to those who love you."

Esau promised, "I will do everything you ask. I will bury you near Sarah when you pass, and I will love Jacob. He is my brother, and it is only natural to love him—we are family. If I do not love him, who else should I love? I also ask that you speak to Jacob about me and my sons. I know he will rule over us, for when my father blessed him, he made him greater and me lesser."

Esau swore to Rebecca that he would do as she asked. Then Rebecca called Jacob to stand before Esau and gave him the same instructions. Jacob replied, "I will do everything you ask. I promise that neither I nor my sons will ever harm Esau. I will only love him."

That evening, they shared a meal and drank together. That night, Rebecca passed away at the age of 155. Esau and Jacob buried her in the cave near Sarah, their grandmother.

Chapter XXXVI.

In the sixth year of that time, Isaac called his two sons, Esau and Jacob, to come to him. He said, "My sons, my time is near. I will soon go to be with our ancestors. When I pass, bury me next to my father Abraham in the cave in the field of Ephron the Hittite—the same tomb Abraham bought for our family. That is where I have prepared my resting place.

I ask you, my sons, to live with honesty and fairness, so that the Lord will fulfill the promises He made to Abraham and his descendants. Love one another as you love your own life. Support each other, work together, and let your bond be strong.

I warn you not to worship idols or be drawn to them. They mislead those who follow them. Remember the Lord, the God of Abraham, and how I served Him with joy and faithfulness. Because of this, He blessed me, made my descendants as numerous as the stars, and established us as a righteous people who will last forever.

Now, I will have you swear a great oath—by the name of the One who created everything in heaven and on earth—that you will honor and worship Him alone. Love your brother with honesty and goodness, and never plan harm against him. If you do this, you will be successful in everything, and no harm will come to you.

But if one of you plots evil against his brother, that person will bring destruction upon himself. He will be cut off from the land of the living, and his family will not survive. When God's wrath comes, as it did in Sodom, his land, city, and everything he owns will be destroyed. His name will be erased from the book of the righteous and placed among those who are condemned. He will face suffering and sorrow forever.

I warn you, my sons: anyone who harms his brother will face judgment. Today, I am dividing my belongings between you. Since Esau is the firstborn, he will receive the larger portion, including the tower and everything Abraham owned at the Well of the Oath."

Esau replied, "I already sold my birthright to Jacob. It belongs to him, and I have no claim to it."

Isaac said, "May God's blessing rest on both of you and your descendants from this day forward. You have brought me peace by making things right between you. My heart is no longer troubled about the birthright. May the Most High bless those who act with righteousness and extend that blessing to their children forever."

After blessing them, Isaac gave his final instructions. They ate and drank together in his presence, and Isaac was filled with joy, knowing his sons were at peace. That night, they went to rest, and Isaac lay in his bed, content. Soon after, he passed away peacefully. He lived to be 180 years old, completing twenty-five weeks and five years. His sons, Esau and Jacob, buried him.

Esau moved to the land of Edom and settled in the mountains of Seir. Jacob remained in the mountains of Hebron, living in the same tower where his grandfather Abraham had once stayed. He continued to worship the Lord with all his heart, following the commands that had been passed down through his family.

In the fourth year of the second week of the forty-fifth jubilee, Jacob's wife Leah passed away. He buried her in the same cave where his mother, Rebecca, had been laid, to the left of Sarah, the mother of his father. All her sons, along with Jacob's other children, came together to mourn her and to comfort Jacob, for he loved her deeply after Rachel had died.

Leah had been kind, faithful, and righteous all her life. She always honored Jacob and never spoke harshly to him. She was gentle, loving, and respectable. Jacob remembered her goodness and mourned her loss with all his heart and soul.

Chapter XXXV II.

On the day Isaac, the father of Jacob and Esau, passed away, Esau's sons found out that Isaac had given the birthright to Jacob, even though Esau was the older son. They were furious and demanded answers from their father.

"Why did your father give the firstborn's blessing to Jacob instead of you?" they asked.

Esau answered, "I sold my birthright to Jacob for a bowl of lentils. Later, when our father sent me to hunt and bring him food so he could bless me, Jacob tricked me. He brought food to our father first and received the blessing that was meant for me. After that, our father made us both swear to live in peace, to love each other, and not to harm one another."

But Esau's sons refused to listen. "We will not make peace with Jacob," they declared. "We are stronger than he is. We will fight him, kill him, and wipe out his sons. And if you refuse to help us, we will deal with you too."

They continued, "Let's send messengers to Aram, Philistia, Moab, and Ammon to gather warriors who love to fight. With their help, we will destroy Jacob before he grows even more powerful."

Esau tried to stop them. "Do not go to war with him," he warned. "If you do, you may be the ones who fall."

But his sons ignored him. "You have spent your whole life obeying Jacob. We will not follow your advice."

Determined to carry out their plan, they sent messengers to Esau's ally, Aduram, and hired a thousand warriors. They also gathered a thousand fighters each from Moab, Ammon, Philistia, Edom, and the Horites. From the Kittim, they brought mighty warriors. Then, turning to Esau, they threatened, "Lead us into battle, or we will kill you."

Filled with anger and frustration, Esau finally agreed. As old feelings of resentment returned, he forgot the oath he had sworn to his parents and allowed his heart to turn against Jacob once again.

Meanwhile, Jacob had no idea that trouble was coming. He was still mourning the death of Leah, his wife, when Esau and his army of four thousand warriors came near the tower where he was staying. The people of Hebron, who respected Jacob more than Esau because of his kindness and generosity, rushed to warn him.

"Esau is coming with four thousand armed men, ready for battle," they told him.

At first, Jacob didn't believe them. But when he saw the army approaching, he quickly shut the gates of the tower and climbed to the top. From there, he called out to Esau.

"Is this how you come to comfort me after my wife's death? Is this how you keep the oath you swore to our father and mother? You have broken your promise and brought judgment upon yourself."

Esau replied bitterly, "Oaths mean nothing. People and animals alike will always fight their enemies. You have hated me and my children for as long as I can remember. We are not brothers, and we never will be."

Then Esau spoke in anger:

"If a wild boar could grow soft fur like a lamb,
    If it could sprout horns like a ram or a deer,
    Then I would make peace with you.

If a mother could leave her newborn child,
    Then I would call you my brother.

If wolves could lie down with lambs,
    Without trying to tear them apart,
    And their hearts became kind,
    Then I would make peace with you.

If a lion and an ox could work together,
> Plowing the fields side by side,
> Then I would make peace with you.

If a raven could turn as white as snow,
> Then you would know that I loved you.

But you and your children will be torn from the land,
> And there will never be peace for you.

When Jacob saw the hatred in Esau's heart and the fury in his eyes, he realized Esau was determined to destroy him. Understanding that words would not change his brother's mind, he compared Esau to a wild boar charging straight into a spear.

Jacob then turned to his men and said, "Prepare your weapons. Stand ready. We will not run from this fight."

Chapter XXXV III.

After that, Judah turned to his father, Jacob, and said, "Father, take your bow and shoot your arrows to defend us. Show your strength, but do not let us harm your brother. He is still your own flesh and blood and should face you in this battle."

Jacob took his bow and fired an arrow, hitting Esau on the right side of his chest, killing him instantly. He shot another arrow, striking 'Adoran the Aramean on his left side, knocking him backward and killing him as well.

Then, Jacob's sons and their servants split into groups and attacked the enemy from all directions around the tower. Judah led the southern group with Naphtali, Gad, and fifty servants. Together, they fought fiercely, leaving no survivors. On the eastern side, Levi, Dan, and

Asher, along with fifty men, faced the warriors from Moab and Ammon, defeating them all. Reuben, Issachar, and Zebulon took the north side with fifty men and overpowered the fighters from Philistia. Meanwhile, Simeon, Benjamin, and Enoch, Reuben's son, attacked from the west with fifty men, defeating four hundred warriors from Edom and the Horites.

Even though six hundred men, including four of Esau's sons, managed to escape, they left Esau's body behind on the hill of 'Aduram. Jacob buried Esau there and returned home.

Jacob's sons pursued Esau's fleeing sons into the mountains of Seir, where they captured them and made them serve Jacob's descendants. They sent word to Jacob, asking if they should make peace with Esau's sons or destroy them completely. Jacob instructed them to make peace, so they did, placing Esau's descendants under their rule and requiring them to pay tribute to Jacob and his sons for generations.

Esau's descendants continued to pay tribute until the day Jacob and his family moved to Egypt. Even today, the people of Edom remain under the rule of Jacob's descendants and have never been able to free themselves from this obligation.

These are the kings who ruled over Edom before Israel had any kings:

- The first king was Balaq, son of Beor, and his city was called Danaba.
- After Balaq died, Jobab, son of Zara from Boser, became king.
- When Jobab died, 'Asam from the land of Teman took the throne.
- After 'Asam's death, 'Adath, son of Barad—who had defeated the Midianites in the field of Moab—became king. His city was called Avith.

- When 'Adath died, Salman from 'Amaseqa ruled as king.
- After Salman, Saul from Ra'aboth by the river took the throne.
- Saul was followed by Ba'elunan, son of Achbor.
- When Ba'elunan died, 'Adath became king again, and his wife was Maitabith, daughter of Matarat, granddaughter of Metabedza'ab.

These kings ruled over Edom before any king was established in Israel.

Chapter XXXI X.

Jacob stayed in the land of Canaan, where his father had once lived. These are the events of his family's story.

Joseph, at seventeen years old, was taken to Egypt and sold to Potiphar, a high-ranking officer of Pharaoh and the captain of the guard. Potiphar put Joseph in charge of his entire household, and because of Joseph, the Lord blessed everything in Potiphar's home. Everything Joseph did was successful, and Potiphar noticed that God was with him, making him prosper in all he did. Because of this, Potiphar gave Joseph complete authority over everything he owned.

Joseph was strong and handsome, which caught the attention of Potiphar's wife. She became obsessed with him and constantly asked him to be with her. But Joseph refused. He remembered what his father, Jacob, had taught him about the words of Abraham: that committing adultery with a married woman was a terrible sin, deserving of death according to God's laws. Joseph held on to these teachings and refused to betray his master or sin against God.

For an entire year, Potiphar's wife tried to convince Joseph, but he resisted. One day, when they were alone, she grabbed his coat and tried to force him to be with her. Joseph pulled away, leaving his coat in her hands, and ran out of the house. Furious at being rejected, she decided to take revenge. She screamed and told the household servants that Joseph had attacked her. Later, when her husband came home, she said, "That Hebrew slave you trust tried to take advantage of me. When I screamed, he ran away, leaving his coat behind."

Hearing this, Potiphar became angry. Seeing the coat as proof, he had Joseph thrown into prison, where the king's prisoners were kept. But even in prison, the Lord was with Joseph. The chief jailer noticed how responsible and successful Joseph was in everything he did. Recognizing that God was with him, the jailer put Joseph in charge of all the prisoners, trusting him completely to manage everything well.

Joseph remained in prison for two years. During that time, Pharaoh became angry with two of his officials—the chief cupbearer and the chief baker—and had them imprisoned in the same place as Joseph. The jailer assigned Joseph to take care of them.

One night, both the cupbearer and the baker had dreams that troubled them. They shared their dreams with Joseph, and with God's help, he explained their meanings. Just as Joseph predicted, the cupbearer was restored to his position, while the baker was executed.

Before the cupbearer was released, Joseph asked him to remember him and mention his situation to Pharaoh, hoping to be freed. However, once the cupbearer returned to his position, he completely forgot about Joseph and did not speak of him to Pharaoh. Despite Joseph's kindness, the cupbearer did not remember him at all.

Chapter XL.

One night, Pharaoh had two dreams about a great famine that was coming to the land. When he woke up, he called all the dream interpreters and magicians in Egypt to explain his dreams. But none of them could understand what they meant. Then, the chief butler remembered Joseph and told Pharaoh about him. Joseph was taken out of prison and brought before Pharaoh to hear the dreams.

Joseph told Pharaoh, "Both of your dreams mean the same thing. There will be seven years of great harvests, with plenty of food across Egypt. But after that, seven years of famine will come, so severe that people will forget the years of plenty."

He then advised Pharaoh, "You should put wise and responsible men in charge of storing extra food in every city during the seven good years. That way, when the famine comes, there will be enough food to keep people alive, and Egypt will not be ruined by hunger."

God gave Joseph wisdom and favor in Pharaoh's eyes. Pharaoh said to his servants, "There is no one as wise as Joseph. The spirit of God is with him." So Pharaoh made Joseph the second most powerful man in all of Egypt. He gave Joseph control over the entire land and had him ride in his second chariot. He dressed Joseph in fine clothes, placed a gold chain around his neck, and gave him his official ring as a sign of his authority. A messenger went before him, announcing his new position. Pharaoh told Joseph, "Only I, as king, will be greater than you."

Joseph ruled fairly over Egypt. The officials and workers respected him, and he treated everyone with honesty. He never took bribes or showed favoritism. Because of him, Egypt was peaceful, and God continued to bless him. People admired Joseph, and Pharaoh's kingdom remained well-run and free from trouble.

Pharaoh gave Joseph a new name, Zaphenath-Paneah, and arranged his marriage to Asenath, the daughter of Potipherah, a priest from On. Joseph was thirty years old when he stood before Pharaoh.

That same year, Isaac passed away. Just as Joseph had predicted, the land had seven years of plenty. The harvests were so abundant that one portion of seed produced eighteen hundred times more. Joseph collected and stored food in every city until the grain supply was so large that it couldn't even be measured.

Chapter XLI.

During the forty-fifth jubilee, in the second week of the second year, Judah arranged for his oldest son, Er, to marry a woman named Tamar, who came from the daughters of Aram. However, Er did not love her because she wasn't from his mother's Canaanite family. He wanted to marry someone from his mother's people, but Judah refused to allow it. Er was wicked in God's eyes, so God took his life.

After Er's death, Judah told his second son, Onan, to marry Tamar and have children on behalf of his late brother. But Onan knew the children wouldn't be considered his, so he purposely avoided making her pregnant. This angered God, and Onan also died.

Judah then told Tamar to stay in her father's house as a widow until his youngest son, Shelah, was old enough to marry her. However, when Shelah grew up, Judah's wife, Bedsu'el, was against the marriage. In the fifth year of that week, Bedsu'el passed away. The next year, Judah went to Timnah to shear his sheep.

When Tamar heard that Judah was going to Timnah, she took off her widow's clothing, covered her face with a veil, and dressed beautifully. She waited by the road where Judah would pass. When Judah saw her, he thought she was a prostitute and approached her. He said, "Let me be with you." Tamar asked, "What will you give me in return?" Judah replied, "I don't have anything with me right now, but I will leave my signet ring, necklace, and staff as a guarantee until I send payment." Tamar agreed, and they were together. As a result, she became pregnant.

Afterward, Tamar returned home. Later, Judah sent a servant with a young goat to pay her and get his items back, but the servant couldn't find her. The locals told him, "There has been no prostitute here." When the servant returned and told Judah, he said, "Let her keep the items. We don't want to be embarrassed."

Three months later, Judah was told, "Tamar, your daughter-in-law, is pregnant from prostitution." Enraged, Judah went to her father's house and demanded that she be burned as punishment. As she was being taken out, Tamar sent a message to Judah, saying, "The man who owns these items is the father of my child. Do you recognize them?" She showed the signet ring, necklace, and staff. Judah immediately knew they were his and admitted, "She is more righteous than I am. I failed to give her to my son Shelah, as I promised." Judah stopped the punishment, and Tamar's life was spared.

Tamar was never married to Shelah, and Judah never had relations with her again. Later, she gave birth to twin boys, Perez and Zerah, in the seventh year of the second week. Around this time, the seven years of abundance that Joseph had predicted for Egypt came to an end.

Judah deeply regretted what he had done. He realized his mistake and sincerely repented before God. Because of his honest confession, God forgave him, and he never repeated the sin. In a dream, he was assured that his wrongdoing had been forgiven because he had shown true remorse and made amends.

It was also revealed to him that his other sons were not responsible for what had happened with Tamar, and because of this, his family line continued. Judah had followed the strict traditions passed down from

Abraham when he first sought to punish Tamar, and his actions later influenced the laws that were established.

Chapter XLII.

In the first year of the third week of the forty-fifth jubilee, a terrible famine spread across the land. The rains stopped, and the ground became dry and lifeless. But in Egypt, there was still food because Joseph had wisely stored grain during the seven good years. As the famine worsened, the Egyptians came to Joseph to buy food, and he opened the storehouses, selling grain in exchange for gold.

Meanwhile, Canaan was struggling. When Jacob heard there was food in Egypt, he sent ten of his sons to buy grain, but he did not let Benjamin go with them. When the brothers arrived in Egypt along with other people looking for food, Joseph recognized them immediately, but they did not know who he was. Acting like a stranger, Joseph accused them of being spies. "Are you here to find weaknesses in our land?" he asked. He then put them in prison for a few days. Later, he released nine of them but kept Simeon as a hostage, telling them to return with their youngest brother to prove they were telling the truth. Without them knowing, Joseph secretly filled their sacks with grain and returned their money.

Back in Canaan, the brothers told Jacob everything that had happened. They explained how the ruler of Egypt accused them of spying and would not release Simeon unless they brought Benjamin back. Jacob was heartbroken. "You have already taken my children from me! Joseph is gone, Simeon is gone, and now you want to take Benjamin too? Everything is against me!" he cried. He refused to let Benjamin go. "His mother had only two sons. One is already gone. If something happens to him, I will be in sorrow for the rest of my life."

When they found their money inside their sacks, they became even more afraid, and Jacob refused to send Benjamin with them. But the famine only grew worse, while Egypt still had plenty of food because of Joseph's careful planning. As their supplies ran low, Jacob told his sons, "Go back and buy more food, or we will starve." But they replied, "We cannot go back unless Benjamin is with us. The man was clear— we must bring him."

Seeing no other choice, Jacob finally agreed. Reuben offered, "Trust him to me. If I don't bring him back, you can take my two sons." But Jacob refused. Then Judah stepped forward and said, "Send him with me. I will take full responsibility. If I don't bring him back, I will carry the blame forever."

At last, Jacob agreed and sent Benjamin with them. He also told them to bring gifts for the Egyptian leader: stacte, almonds, terebinth nuts, and pure honey. On the first day of the second year of the famine, they left for Egypt.

When they arrived, Joseph immediately recognized Benjamin but did not reveal who he was. "Is this your youngest brother?" he asked. "Yes," they answered. Joseph then said, "May the Lord be gracious to you, my son."

Joseph invited them to his house, released Simeon, and prepared a feast. The brothers gave him the gifts, and they all ate and drank together. During the meal, Joseph gave food to each of them, but he gave Benjamin seven times more than the others.

Before they left, Joseph wanted to test them. He told his servant, "Fill their sacks with grain, return their money, and put my silver cup— the one I drink from—in the youngest brother's sack. Then send them on their way."

Chapter XLIII.

The steward followed Joseph's instructions exactly. He filled the brothers' sacks with food, returned their money, and secretly placed Joseph's silver cup in Benjamin's sack. Early the next morning, the brothers left for home. But soon after, Joseph told his steward, "Go after them. When you catch them, ask, 'Why have you repaid kindness with betrayal? You stole my master's special cup!' Bring the youngest back to me right away—I must decide what to do with him."

The steward chased after them, caught up, and repeated Joseph's words. The brothers were shocked and said, "Why would you accuse us of this? We would never steal from your master! We even brought back the money we found in our sacks last time. Why would we take silver from his house? Search our bags! If you find the cup with anyone, let him die, and the rest of us will become your slaves."

The steward replied, "No, only the one who has the cup will stay as my servant. The rest of you may go free."

He searched their sacks, starting with the oldest and ending with the youngest. When he opened Benjamin's sack, there was the silver cup. The brothers were devastated. They tore their clothes in grief, loaded their donkeys, and returned to the city.

When they reached Joseph's house, they fell to the ground before him. Joseph asked, "What have you done? Didn't you think I would find out the truth?"

The brothers answered, "What can we say? How can we prove our innocence? God has exposed our guilt. We are your servants now, including the one who had the cup."

Joseph responded, "I fear God. I will not punish all of you. Only the one who stole the cup will stay as my servant. The rest of you may return to your father."

At this, Judah stepped forward and pleaded, "My lord, please listen. We have an elderly father who loves his youngest son dearly. His life is tied to this boy. If we return without him, our father will die of heartbreak. Please let me stay as your servant instead, and let the boy go home with his brothers. I promised my father I would bring him back safely. If I fail, I will carry this guilt forever."

Joseph could no longer hold back his emotions. Seeing their love for one another, he ordered everyone else to leave the room. Then, with tears streaming down his face, he said in Hebrew, "I am Joseph, your brother."

The brothers were too shocked to speak. Joseph continued, "It's really me—the one you sold into Egypt. But don't be afraid or blame yourselves. God sent me here ahead of you to save lives. This famine has already lasted two years, and there are still five more years without harvests or food. God used me to ensure survival. Hurry back to our father and tell him I'm alive and that God has made me ruler of all Egypt. Bring him and your families here so I can take care of you during the remaining years of famine."

Joseph hugged each of his brothers, crying with them. Then he provided them with wagons, supplies for their journey, fine clothes, and silver. For his father, he sent ten donkeys loaded with the best goods from Egypt, along with grain and bread for the trip.

When the brothers returned to Canaan, they told their father, "Joseph is alive! He is the ruler of all Egypt!" Jacob was stunned and couldn't believe it at first. But when he saw the wagons and all the provisions Joseph had sent, his spirit was lifted. He said, "That is enough! My son Joseph is alive! I will go and see him before I die."

Chapter XLIV.

Israel left Haran on the first day of the third month, beginning his journey to Egypt. On the seventh day, he reached the Well of the Oath and offered a sacrifice to the God of his father, Isaac. As he remembered the dream he had at Bethel, he felt unsure and afraid to continue to Egypt. He thought about sending for Joseph instead, so he wouldn't have to leave Canaan. He stayed there for seven days, seeking guidance and hoping for a sign. During this time, he observed the festival of the first-fruits, even though there was no grain to plant because of the severe famine, which had affected crops, animals, birds, and people alike.

On the sixteenth day of the month, the Lord appeared to Jacob in a vision and called, "Jacob, Jacob." Jacob answered, "Here I am." The Lord said, "I am the God of your fathers, the God of Abraham and Isaac. Do not be afraid to go to Egypt. I will make your family into a great nation there. I will be with you and bring you back safely. You will be buried in your homeland, and Joseph will be with you when you die. Have no fear—go to Egypt."

Encouraged by this, Jacob gathered his sons, grandsons, and belongings. They placed him and all they owned on wagons, and on the sixteenth day of the third month, they left the Well of the Oath. Judah went ahead to meet Joseph and prepare the land of Goshen, which Joseph had chosen as their new home. Goshen was a good place for them because it was fertile and close to Joseph, making it ideal for their livestock and families.

These were the family members who traveled with Jacob to Egypt:

- Reuben, Jacob's firstborn, and his sons: Enoch, Pallu, Hezron, and Carmi—four in total.
- Simeon and his sons: Jemuel, Jamin, Ohad, Jachin, Zohar, and Shaul (whose mother was a Zephathite woman)—seven.
- Levi and his sons: Gershon, Kohath, and Merari—three.
- Judah and his sons: Shela, Perez, and Zerah—three.
- Issachar and his sons: Tola, Phua, Jasub, and Shimron—four.
- Zebulun and his sons: Sered, Elon, and Jahleel—three.

These were the descendants of Leah, along with their sister Dinah, who were born to Jacob in Mesopotamia. Including Jacob himself, thirty members of Leah's family entered Egypt.

From Zilpah, Leah's maidservant:

- Gad and his sons: Ziphion, Haggi, Shuni, Ezbon, Eri, Areli, and Arodi—seven.
- Asher and his sons: Imnah, Ishvah, Ishvi, Beriah, and their sister Serah—six.

Zilpah's descendants who traveled to Egypt numbered fourteen.

From Rachel, Jacob's beloved wife:

- Joseph, who had two sons in Egypt, Manasseh and Ephraim (born to Asenath, daughter of Potiphar, priest of Heliopolis)—two.
- Benjamin and his ten sons: Bela, Becher, Ashbel, Gera, Naaman, Ehi, Rosh, Muppim, Huppim, and Ard—ten.

Rachel's descendants who entered Egypt totaled fourteen.

From Bilhah, Rachel's maidservant:

- Dan and his sons: Hushim, Samon, Asudi, Ijaka, and Salomon—five (though only Hushim survived after arriving in Egypt).
- Naphtali and his sons: Jahziel, Guni, Jezer, Shallum, and Iv— five (Iv was born after the famine years but passed away in Egypt).

Bilhah's descendants totaled twenty-six.

In total, seventy of Jacob's family members traveled to Egypt, including his children and grandchildren. However, five of them— Judah's two sons, Er and Onan, and three others who died without children in Egypt—were buried and counted among the seventy nations of the world.

Chapter XLV.

Israel arrived in Egypt and settled in Goshen on the first day of the fourth month in the second year of the third week of the forty-fifth jubilee. Joseph traveled to Goshen to greet his father, and when they met, he hugged Jacob tightly and wept on his shoulder. Israel said, "Now I can die in peace, because I have seen your face and know that you are alive. Praise the Lord, the God of Israel, the God of Abraham and Isaac, who has shown me mercy and kept His promises. It is enough for me that I have seen you. The vision I had at Bethel has come true. Blessed be the Lord, my God, forever and ever."

Joseph and his brothers sat down and ate together in Jacob's presence. Seeing them reunited, sharing a meal, filled Jacob with great happiness. He thanked God, who had watched over him and kept all twelve of his sons safe.

Joseph arranged for his father, brothers, and their families to live in Goshen, specifically in Rameses and the nearby areas, which were under his control as Pharaoh's ruler. Israel and his family settled in the most fertile part of Egypt. Jacob was 130 years old when he arrived, and Joseph made sure they had enough food throughout the remaining years of famine.

As the famine continued, Joseph collected all the land in Egypt for Pharaoh in exchange for food. He also took the people's livestock and possessions for Pharaoh. When the famine finally ended, Joseph provided the Egyptians with seed in the eighth year so they could plant again. The Nile had finally overflowed its banks, marking the end of the food shortage. During the seven years of famine, the river had failed to flood, only watering the edges of the land. But now, it once again covered the fields, allowing the Egyptians to grow crops and harvest an abundance that year. This was the first year of the fourth week of the forty-fifth jubilee.

Joseph established a law in Egypt that required one-fifth of all harvests to go to Pharaoh, while the remaining four-fifths were for the people to use as food and seed. This law remained in place for generations.

Israel lived in Egypt for seventeen more years, reaching a total age of 147 years, or three jubilees. He passed away in the fourth year of the fifth week of the forty-fifth jubilee. Before he died, he gathered his sons, blessed them, and told them what would happen in Egypt and in the future. He gave each of them a blessing and granted Joseph a double portion of inheritance in the land.

Israel was buried in the double cave in Canaan, near his father Abraham, in the tomb that Abraham had prepared in Hebron. Before his death, Israel gave all his writings and the books of his ancestors to Levi, instructing him to protect them and pass them down through future generations so they would never be lost.

Chapter XLVI.

After Jacob passed away, his descendants thrived in Egypt. Their families grew quickly, and they became a large, united people. The brothers cared for one another, and everyone worked together to support their community. During Joseph's lifetime, they increased greatly in number over ten cycles of seven years. There was no trouble or conflict because the Egyptians respected and valued the Israelites while Joseph was alive.

Joseph lived to be 110 years old. He spent 17 years in Canaan, 10 years as a servant, 3 years in prison, and 80 years as a ruler in Egypt. Before he died, he made the Israelites promise that when they eventually left Egypt, they would take his bones with them. He knew the Egyptians would not allow him to be buried in Canaan.

This was because King Makamaron of Canaan, who was living in Assyria at the time, had fought against the Egyptian king in a valley and defeated him, forcing him to retreat to the gates of 'Ermon. However, Makamaron was unable to enter Egypt because a new, stronger ruler had taken power. The gates of Egypt were heavily guarded, and no one was allowed to pass through.

Joseph died in the forty-sixth jubilee, during the sixth week, in the second year, and he was buried in Egypt. Eventually, all of his brothers passed away as well, along with their entire generation.

In the forty-seventh jubilee, during the second week of the second year, the king of Egypt went to war against the king of Canaan. Around this time, the Israelites took the remains of Jacob's sons, except for Joseph's, and buried them in the double cave on the mountain. Most of the Israelites returned to Egypt, but a few stayed in the mountains of Hebron, including Amram, your father, who remained with them.

Later, the king of Canaan defeated the Egyptian king and sealed off Egypt's borders. Afterward, he created a harsh plan against the Israelites. He told his people, "The Israelites have grown too large and strong. We must act now before they increase even more. If war breaks out, they might join our enemies and leave Egypt. Their hearts are already set on returning to Canaan."

To control them, he put slave masters over them and forced them to build strong cities for Pharaoh, including Pithom and Raamses. They were also made to repair and strengthen Egypt's cities. The Israelites were treated cruelly, but the more they were oppressed, the more their numbers grew. This made the Egyptians fear and resent them even more, leading to even harsher treatment.

Chapter XLVII.

During the seventh week, in the seventh year of the forty-seventh jubilee, your father left Canaan. You were born in the fourth week, during the sixth year of the forty-eighth jubilee. At that time, the Israelites were suffering greatly. Pharaoh, the ruler of Egypt, had ordered that all newborn Hebrew boys be thrown into the river. For seven months, this cruel law was strictly followed, and many baby boys were cast into the waters.

On the day you were born, your mother hid you for three months to keep you safe. When she could no longer hide you, she made a small basket, sealing it with pitch and tar so it would float. She placed you inside and set it among the reeds along the riverbank. For seven days, she returned at night to nurse you, while your sister Miriam stayed nearby during the day to watch over you and protect you from harm.

One day, Pharaoh's daughter, Tharmuth, came to the river to bathe. She heard your cries and told her maids to bring her the basket. When she saw you inside, she felt compassion and decided to adopt you. Your sister stepped forward and asked, "Shall I find a Hebrew woman to nurse the baby for you?" Tharmuth agreed, and Miriam brought your mother, Jochebed, to care for you. Pharaoh's daughter even paid her to look after you.

When you grew older, your mother brought you back to Tharmuth, who raised you as her own son. Although you were brought up in Pharaoh's palace, your father, Amram, secretly taught you how to read and write, making sure you knew your true heritage. You spent 21 years in the royal court, but one event changed your life forever.

One day, while walking outside the palace, you saw an Egyptian beating one of your fellow Israelites. Overcome with anger, you killed the Egyptian and buried his body in the sand to cover it up. The next day, you saw two Israelites arguing and tried to stop them. You asked the one in the wrong, "Why are you hitting your brother?" But he pushed back and said, "Who made you our ruler or judge? Are you planning to kill me like you killed the Egyptian?"

When you heard this, fear took hold of you. You realized that people knew what you had done, and it would only be a matter of time before Pharaoh found out. Worried for your safety, you fled Egypt to escape the consequences of your actions.

Chapter XLVIII.

In the sixth year of the third week of the forty-ninth jubilee, you left and lived in Midian for five weeks and one year. Then, in the second week of the second year of the fiftieth jubilee, you returned to Egypt. You clearly remember what God told you on Mount Sinai and how Prince Mastêmâ tried to stop you on your way back. He saw that you had been sent to bring judgment on Egypt and used all his power to try and kill you, hoping to prevent you from saving the Israelites. But I rescued you from his grasp, and you carried out the signs and wonders that God had commanded you to perform in Egypt against Pharaoh, his household, his officials, and his people.

The Lord sent powerful judgments against the Egyptians for the sake of Israel. He struck them with plagues: turning the water to blood, covering the land with frogs, tormenting them with lice and gnats, afflicting them with painful boils, killing their livestock, sending hail that destroyed their crops, unleashing locusts that ate whatever was left, covering the land in darkness, and finally, taking the lives of all their firstborn children and animals. The Lord also destroyed their idols, burning them with fire.

Everything happened exactly as you foretold. In front of Pharaoh, his officials, and all of Egypt, you warned them, and each plague came just as you had said. The Lord sent ten devastating plagues to punish Egypt and avenge Israel. He did this to keep His promise to Abraham, repaying the Egyptians for enslaving His people.

But Prince Mastêmâ fought against you the entire time. He tried to hand you over to Pharaoh and helped the Egyptian magicians perform evil tricks. However, we prevented their magic from healing anyone. Instead, the Lord struck them with painful sores so severe that they couldn't even stand, preventing them from performing any more illusions.

Even after witnessing all these miracles, Mastêmâ did not give up. Instead, he encouraged the Egyptians to chase after the Israelites with their full army—chariots, horses, and soldiers. But I stood between them and Israel, protecting My people and delivering them from his hands.

The Lord led Israel safely through the sea, turning the water into dry land so they could cross. But when the Egyptians followed, the Lord threw them into the deep waters, drowning them. Just as they had drowned Israelite children in the river, God repaid them, destroying one million of them, wiping out a thousand strong men for every Hebrew child they had thrown into the water.

On the fourteenth, fifteenth, sixteenth, seventeenth, and eighteenth days, Mastêmâ was bound and held back so he could not accuse the Israelites. On the nineteenth day, we released him, allowing him to influence the Egyptians as they pursued Israel. But God hardened their hearts, making them even more stubborn. This was part of His plan—to bring them to their destruction in the sea.

On the fourteenth day, we bound Mastêmâ so he could not accuse Israel when they took gold, silver, bronze, and clothing from the Egyptians. This was their rightful payment for all the years they had been forced to work as slaves. The Lord made sure the Israelites did not leave Egypt empty-handed.

Chapter XLIX.

Remember the command the Lord gave you about Passover: celebrate it at the right time, on the fourteenth day of the first month. The sacrifice must be made before evening and eaten that same night, as the fifteenth day begins at sunset.

On this special night—when the festival begins and joy fills the hearts of Israel—you were eating the Passover meal in Egypt. That same night, all the forces of Mastêmâ were released to strike down every firstborn in Egypt, from Pharaoh's son to the firstborn of the lowest servant, even the firstborn animals.

The Lord gave His people a sign: any house with the blood of a one-year-old lamb on its doorframe would be protected. The destroyer would not enter but would pass over, sparing everyone inside because of the blood on the door.

The Lord's power worked exactly as He commanded. The plague passed over the houses of the Israelites, leaving them unharmed. No person, animal, or even a dog suffered any harm. But in Egypt, the disaster was severe—every household lost someone, filling the land with mourning and cries of sorrow.

Meanwhile, the Israelites were eating the Passover lamb, drinking wine, and giving thanks to the Lord, praising Him for their deliverance. They were ready to leave Egypt and escape from their suffering.

Remember this day always, and celebrate it every year on the appointed day, following all the instructions. Do not delay or change the date.

This is an everlasting command, recorded in the heavenly books. Every generation of Israel must observe it every year, on the exact date. This law will never change.

Anyone who is able but refuses to celebrate the Passover on the correct day—failing to offer a sacrifice to the Lord and join in the feast—will be cut off from Israel. Because they ignored the Lord's command, they will bear the guilt of disobedience.

The people of Israel must celebrate Passover on the fourteenth day of the first month, from evening to evening, as the day transitions from light to night. The Lord has commanded that it be observed at this specific time, "between the evenings."

The Passover sacrifice must not be made during the day but only at sunset. It must be eaten that night until the first third of the night has passed. Any leftover meat must be burned.

The lamb must not be boiled or eaten raw. It must be roasted over fire with its head, insides, and legs intact. Its bones must not be broken, for just as no Israelite's bones shall be broken, neither shall the bones of the Passover lamb.

The Lord commanded Israel to observe this festival on its exact date, without postponing it. It is a holy day, a time set apart for worship, and must not be rescheduled or moved.

Tell the people of Israel to celebrate Passover every year, as the Lord commanded. It will be a lasting memorial that pleases Him, and no plague will harm those who keep this command.

The Passover meal must not be eaten outside the Lord's sanctuary. All of Israel must come together and celebrate it at the right time.

Every man who is at least twenty years old on the day of Passover must eat it in the Lord's sanctuary, as it is written. They must partake in the feast before the Lord.

When the Israelites enter the land of Canaan—the land given to them as their inheritance—and establish the Lord's tabernacle in the center of the land, in one of their tribes, they must continue celebrating Passover at the tabernacle every year. They must sacrifice the lamb before the Lord as part of their worship.

When the Lord's temple is built in the land, the people must go there to offer the Passover sacrifice at sunset. The lamb's blood must be placed at the altar's entrance, its fat burned on the altar fire, and its meat roasted and eaten in the courtyard of the holy temple.

Passover must not be celebrated in private homes or in different cities. It must only be observed at the tabernacle or the temple where the Lord's name dwells. The people must remain faithful and not turn away from Him.

Moses, instruct the Israelites to follow these Passover commands exactly as I have given them to you. Teach them to celebrate this festival each year and observe the Feast of Unleavened Bread for seven days. During these seven days, they must eat unleavened bread and bring daily offerings before the Lord at His altar.

This festival is a reminder of the night you left Egypt in haste and entered the wilderness of Shur. You completed the celebration by the sea.

Chapter L.

After giving you this law, I also told you about the Sabbath days while you were in the desert of Sin, between Elim and Sinai. I explained the Sabbaths for the land on Mount Sinai, and I also told you about the cycle of jubilee years. However, I did not tell you about the year of the jubilee yet because you will only observe it after entering the land that you will possess. The land itself will also observe the Sabbaths while you live in it, and then you will understand the jubilee year.

For this reason, I have established for you the system of weeks, years, and jubilees. There have been forty-nine jubilees from the time of Adam until today, plus one week and two years. There are still forty more years left for you to learn the commandments of the Lord before you cross over the Jordan River into the land of Canaan. The cycle of jubilees will continue until Israel is completely purified from sin, wrongdoing, and impurity. When that time comes, Israel will live in peace, free from Satan and all evil, and the land will remain pure forever.

I have written down the commandment about the Sabbaths for you, along with all the rules and judgments that come with it. You are to work for six days, but the seventh day is the Sabbath of the Lord your God. On this day, no one should work—not you, your children, your servants, your animals, or any visitor staying with you. Anyone who works on the Sabbath must be put to death.

Anyone who dishonors the Sabbath in any way—by engaging in intimate relations, planning work, starting a journey, buying or selling, drawing water that was not prepared on the sixth day, or carrying a load from their house—must also be put to death. You are to do no work on the Sabbath except what was prepared in advance for eating, drinking, and resting. The Sabbath is a day to bless the Lord, who has given you a special and holy day of rest. It is a day for all of Israel to stop working and observe forever.

The Lord has honored Israel by allowing them to eat, drink, and rest on this festival day, free from the labor of men. The only work that should be done on the Sabbath is offering incense, sacrifices, and offerings to the Lord. These acts of worship must take place in the Lord's sanctuary so that atonement can be made for Israel as a lasting memorial that pleases God. These offerings should be presented to Him daily, as He has commanded.

Anyone who works on the Sabbath, travels, tends to their farm, lights a fire, rides an animal, sails a boat, harms or kills any creature, slaughters an animal or bird, catches any fish, fasts, or goes to war on this day must be put to death. This is so the people of Israel will keep the Sabbath properly, as written in the commandments given to me. These laws were recorded on the tablets, teaching the people about the seasons and how to observe their days.

This completes the instructions about how the days are divided.

Temple Scroll

Introduction

Among the extraordinary discoveries made in the Judean Desert near the Dead Sea, none have sparked as much scholarly fascination and theological reflection as the Temple Scroll. Unearthed from the renowned caves of Qumran, where the Dead Sea Scrolls lay hidden for nearly two millennia, the Temple Scroll offers a captivating glimpse into ancient Jewish spirituality, religious law, and eschatological expectations. This remarkable text stands apart not only because of its physical characteristics—being the longest scroll discovered among the Dead Sea manuscripts—but also because of its content, which meticulously describes the ideal temple, its rituals, priestly practices, and the societal organization envisioned by its anonymous author.

The Temple Scroll is far more than merely an archaeological curiosity or a historical artifact; it represents a theological manifesto of a devout community, a visionary blueprint for an idealized religious society, and a profound commentary on the biblical tradition. It synthesizes diverse elements of Israelite religious thought and practice, providing a unique window into the dynamic and complex religious landscape of Second Temple Judaism. To read the Temple Scroll is to step into the religious imagination of an ancient community deeply concerned with purity, holiness, and the proper worship of God.

Despite its significance, the Temple Scroll remains relatively unfamiliar to many contemporary readers outside specialized scholarly circles. Yet its detailed depiction of temple architecture, sacrificial rituals, purity laws, and social organization offers essential insights into the development of Jewish religious practices and the origins of later rabbinic traditions. Moreover, the scroll provides crucial context for understanding the environment from which Christianity emerged, illuminating aspects of religious belief and practice that influenced both Judaism and early Christian communities.

This comprehensive introduction aims to guide readers deeply into the rich tapestry of historical context, literary composition, and theological themes of the Temple Scroll. By providing clarity and context, readers will be better equipped to explore and appreciate the profound significance and enduring relevance of this ancient text.

Historical Background and Discovery

The Temple Scroll was discovered during one of the most extraordinary archaeological events of the 20th century—the discovery of the Dead Sea Scrolls. Found in a cave near Qumran (designated Cave 11) in 1956, the scroll quickly gained attention due to its exceptional length, preservation, and distinctiveness from other documents in the collection. Unlike most scrolls at Qumran, which are largely fragmentary, the Temple Scroll was remarkably intact, measuring approximately 8.75 meters (29 feet) in length, making it the longest of all the Dead Sea Scrolls. Its substantial size reflects the immense importance attributed to it by the community responsible for its preservation.

Scholars date the composition of the Temple Scroll to roughly between the late 2nd century BCE and the early 1st century BCE, a period marked by considerable turmoil and religious transformation within Jewish society. This era, known historically as the Hasmonean period, was characterized by significant religious reforms, political struggles, and heightened expectations for the restoration and renewal of Israelite worship. The scroll's composition coincided with profound debates about temple authority, priestly legitimacy, and purity laws—central issues that would deeply shape Judaism during and after the Second Temple period.

The identity of the community responsible for the Temple Scroll's composition and preservation remains a matter of scholarly debate, although the majority of scholars attribute it to the Essenes, a Jewish sect that lived near the Dead Sea in Qumran. The Essenes are traditionally understood as a separatist community devoted to rigorous adherence to purity laws and the expectation of an imminent eschatological age. Their strict interpretation of scripture and rejection of the Jerusalem Temple authorities are evident in numerous Dead Sea Scrolls, but nowhere as clearly articulated as in the Temple Scroll. This document vividly illustrates the community's vision of an idealized society centered around a perfect temple—a vision that stood in stark contrast to contemporary religious practices in Jerusalem.

Historical circumstances also explain why the scroll was hidden. As Roman forces advanced during the Jewish Revolt (66–73 CE), the community likely concealed their sacred texts in nearby caves to protect them from destruction. There they remained, undiscovered for almost two millennia, preserving the theological and cultural aspirations of a community whose historical experience was marked by profound spiritual intensity and fervent hope for divine intervention and restoration.

Literary Composition and Thematic Structure

At first glance, the Temple Scroll presents itself as an authoritative divine revelation, styled directly as the speech of God to Moses. This literary technique—known as pseudepigraphy—was a common practice in Second Temple Judaism, lending profound authority and legitimacy to the document. The scroll extensively quotes and interprets material from biblical sources, particularly from Exodus, Leviticus, Numbers, and Deuteronomy, while significantly expanding upon and modifying these texts. Thus, the Temple Scroll is not merely a repetition of scripture but a creative reworking of biblical law and temple ritual intended to reflect an idealized version of Israel's worship and societal structure.

The scroll begins with meticulous instructions concerning the construction and layout of an ideal temple complex, far surpassing the descriptions found in biblical texts. The temple described is larger, more elaborate, and architecturally ambitious than any historical temple in Jerusalem, signifying the scroll's aspiration toward perfection and holiness. This detailed temple plan symbolizes the belief that true worship must occur within an environment perfectly suited to divine standards. Its elaborate descriptions emphasize spatial holiness, carefully delineating sacred and profane areas, reflecting the Essene community's obsession with purity and separation from impurity.

Following the architectural descriptions, the scroll provides extensive legislation governing sacrifices, festivals, purity regulations, priestly duties, and even civil administration. The level of detail found here is remarkable, reflecting the author's intent to establish a society meticulously ordered around divine commandments. Ritual purity and moral holiness are central themes, underscoring the community's emphasis on rigorous adherence to religious law as essential to maintaining a relationship with God.

One particularly striking aspect of the scroll is its rigorous enforcement of purity laws, which extend far beyond those specified in biblical sources. The Temple Scroll presents detailed regulations on issues ranging from the purity of individuals, food, and garments, to ritual practices required for purification. These rigorous standards reveal a profound theological conviction: maintaining purity was not merely symbolic but a practical necessity for divine favor and communal sanctity.

Another notable theme of the scroll is its social legislation, including detailed instructions on judicial procedures, kingship regulations, and communal ethics. This legislative dimension of the scroll indicates a strong desire for a comprehensive societal structure firmly rooted in divine law. The scroll thus represents not merely a religious document but a blueprint for an entire way of life, governed by principles believed to originate directly from divine authority.

Enduring Significance and Contemporary Relevance

Although the Temple Scroll originates from an ancient historical context vastly different from modern life, its significance and relevance persist today, particularly for those interested in religious history, theology, and biblical interpretation. Its detailed depictions of temple worship and ritual provide essential insights into ancient Judaism, deepening our understanding of how religious practices evolved over time. Moreover, the scroll offers crucial context for understanding the historical backdrop against which Christianity and rabbinic Judaism developed, illuminating common theological threads and differences between these traditions.

For modern readers, the Temple Scroll challenges contemporary notions of worship, holiness, and religious law. Its emphasis on purity and separation invites reflection on modern practices of spirituality, ethics, and religious observance. The rigorous standards outlined in the scroll may seem severe or overly restrictive, yet they highlight a profound commitment to living in alignment with spiritual ideals—a commitment that can inspire contemporary communities toward greater mindfulness in their spiritual practice.

Additionally, the scroll's detailed vision of an idealized society and worship environment resonates deeply in an age marked by religious pluralism, social fragmentation, and ethical ambiguity. Its aspirations toward societal perfection, governed by divine law and communal responsibility, can encourage critical reflection on contemporary issues related to social justice, community cohesion, and ethical accountability.

In sum, the Temple Scroll is more than an ancient document; it is a testament to human aspiration toward holiness, justice, and spiritual purity. As you engage with its extraordinary visions and meticulous legislation, may you find not only historical understanding but also spiritual insight, inspiring deeper reflection on your own beliefs and the enduring quest for divine presence and ethical living in every age.

Temple Scroll

II [Behold, I will make a covenant.] [For it is something dreadful that I] will do [to you.] [I myself will expel from before you] the A[morites, the Canaanites, the Hittites, the Girgashit]es, the Pe[rizzites, the Hivites and] the Jebusites. Ta[ke care not to make a cove]nant with the inhabitants of the country [which you are to] enter so that they may not prove a sn[are for you.] You must destroy their [alta]rs, [smash their] pillars [and] cut down their [sacred trees and burn] [their] idols [with fire]. You must not desire silver and gold so [that you may not be ensnared by them; for that would be abominable to me]. You must [not] br[ing any abominable idol] into your house [and come] under the ban together with it. You shall de[test and abominate it], for it is under the ban. You shall not worship [another] go[d, for YHWH, whose name is] [Jealous], is a jealous God. Take care not to make a [covenant with the inhabitants of the country] [so that, when they whore] after [their go]ds [and] sacrifice to [them and invite you,] [you may not eat of their sacrifices and] t[ake their daughters for your sons, and their daughters may not whore after] their [gods] and cau[se your sons to whore after them.] ... 94 XIII [This is what you shall offer on the altar:] t[wo y]ear[ling lambs] without blemish [every day as a perpetual holocaust. You shall offer the first in the morning; and you shall offer the other lamb in the evening; the corresponding grain-offering will be a te]nth of fine flour mixed with [a quarter of a hin of beaten oil; it shall be a perpetual holocaust of soothing odour, an offering by fire] to YHWH; and the corresponding drink-offering shall be a quart[er of a hin of] wine. [The priest who offers the holocaust shall receive the skin of] the burnt-[offering which he has offered. You shall offer the other lamb in the even]ing with the same grain-[offering as in the] morning and with the corresponding drink-offering as an offering by fire, a soothing odour to YHWH ... On the S[abbath] days you shall offer two [yearling rams without blemish and two] XIV [tenths of an ephah of fine flour, mixed with oil, for a grain-offering and the corresponding drink-offering. This is the holocaust of every Sabbath in addition to the perpetual holocaust and the corresponding drink-offering. On the first day of each month you shall offer a holocaust to YHWH: two young bulls, one ram, seven yearling rams without blemish and a grain-offe]ring of fine flour, [three tenths of an ephah] mix[ed with half a hin of oil, and a drink-offering, ha]lf a hin for [each young bull and a grain-offering of fine flour mixed with oil, two tenths of an ephah] with a third [of a hin, and wine for a drink-offering, one third of a hinfor each ram;] ... one tenth [of fine flour for] a grain-[offering, mixed with a quarter of a hin, and wine, a quarter of a hi]n for each lamb ... a soothing [odour] to YHWH on the first day of each month. This is the burnt-offering for each month for the months of the year ... On the first day of the [first] month [the months (of the year) shall start; it shall be the first month] of the year [for you.

You shall do no] work. [You shall offer a he-goat for a sin-offering.] It shall be offered by itself to expiate [for you. You shall offer a holocaust: a bullock], a ram, [seven yearli]ng ram lambs [without blemish] ... [ad]di[tional to the bu]r[nt-offering for the new moon, and a grainoffering of three tenths of fine flour mixed with oil], half a hin [for each bullock, and wi]ne for a drink-offering, [half a hin, a soothing odour to YHWH, and two] tenths of fine flour mixed [with oil, one third of a hin. You shall offer wine for a drink-offering,] one th[ird] of a hin for the ram, [an offering by fire, of soothing odour to YHWH; and one tenth of fine flour], a grain-offerin[g mixed with a quarter of a hinof oil. You shall offer wine for a drink-offering, a quarter of a hin] for each [ram] ... lambs and for the he-g[oat] ... XV [ea]ch day ... seven [year]ling [lambs] and a he-[goat] ... according to this statute. For the ordination (of the priests), one ram for each [day, and] baskets of bread for all the ra[ms of the ordination, one basket for] each [ram]. They shall divide all the rams and the baskets for the seve[n days of the ordination for each] day; according to [their] division[s, they shall offer to YHWH the right thigh] of the ram as a holocaust and [the fat covering the entrails and the] two kidneys and the fat on them [and on] the loins and the whole fat tail close to the backbone and the appendage of the liver and the corresponding grain-offering and drink-offering according to the sta[tute. They shall take one unleavened cake from the] basket and one cake of bread with oil and [one] wafer, [and they shall put it all on the fat] together with the offering of the right thigh. Those who sacrifice shall wave the rams and the baskets of bread as a wa[ve-offering be]fore YHWH. This is a holocaust, an offering by fire, of soothing odour before YHWH. [They shall burn everything on the altar over] the holocaust, to complete their ordination during the seven days of [ordination]. If the High Priest is to [minister to YHWH, whoever] has been ordained to put on the vestments in place of his father, shall offer [a bull fo]r all the people and another for the priests.

He shall offer the one for the priests first. The elders of the priest[s] shall lay [their hands] XVI [on] its [hea]d and after them the High Priest and all the [priests. They shall slaughter] the bull [before YHWH]. The elders of the priests shall take from the blood of the bull and [place] it [with their finger on the horns of the altar] and they shall pour [the blood] around the four corners of the [altar] ledge ... [and they shall take from its blood and pl]ace it [on his right ear lobe and on the thumb of his right hand and the big toe of his] right [foot. They shall sprinkle on him and his vestments some of the blood which was on the altar]... [he] shall be [holy] all his days. [He shall not go near any dead body]. He shall [not] render himself unclean [even for his father or mother,] for [he is] hol[y to YHWH, his God] ... [He shall offer on the al]tar and burn [the fat of the first bull] ... [all] the fat on the entrails and [the appendage of the liver and the two kidne]ys and the fat on the[m] and [the fat on] the loins, and the corresponding grain-offering and drink-[offering according to their statute,] he shall bur[n them on the altar.] It shall be [a burnt-]offering, an offering by fire, of soothing odour be[fore YHWH. The flesh of the bull], its skin and offal, they shall burn outside the [sanctuary city on a wood fire] in a place reserved for sin-offerings. There they shall bur[n it with its head and legs] together with all its entrails. They shall burn all of it there except the fat. It is a sin-[offering]. He shall take the second bull, which is for the people, and by it he shall expiate [for all the people of] the assembly, by its blood and fat. As he did with the fir[st] bull, [so he shall do] with the bull of the assembly. He shall place with his finger some of its blood on the horns of the [altar, and the remainder of] its blood, he shall sprinkle o[n the f]our corners of the altar ledge, and [its fat and] the corresponding [grain-] offering and drink-offering, he shall burn on the altar. It is a sin-offering for the assembly. XVII ... They shall rejoice because expiation has been made for them ... This day [shall] be a holy gathering for them, [an eternal rule for all their generations] wherever they dwell. They shall rejoice and ... [Let] them [prepare on the fourtee]nth day of the first month [between dusk and dark the Passover of YHWH]. They shall sacrifice (it) before the evening offering and shall sacrifice ... men from twenty years of age and over shall prepare it. They shall eat it at night in the holy courts. They shall rise early and each shall go to his tent ... On the fifteenth day of this month (there shall be) a ho[ly] gathering. You shall do no work of labour on it. (It shall be) a seven-day feast of unleavened bread for YHWH. You shall offer on each of the[se] seven days a holocaust to YHWH: two young bulls, a ram, and seven ram lambs without blemish and a he-goat for a sin-offering and the corresponding grain-offering and drink-offering [according to the sta]tute for the young bulls, rams, l[am]bs and the he-goat. On the seventh day [(there shall be) an assembly] for [YHWH]. You shall do no work on it. XVIII ... [he-] goat for a sin-offering ... [the corresponding grain-offering and drink-] offering according to the statute; one tenth of fine flour [mixed with a quarter of a hin of oil and] a quarter of a hin of wine for a drink-offering ... [he shall expiate] for all the guilt of the people of the assembly ... This shall be an eternal [ru]le for you [for your generations wherever you dwell.] Then they shall offer the one ram, on[ce], on the day of the waving of the sheaf. You shall count seven complete Sabbaths from the day of your bringing the sheaf of [the wave-offering. You shall c]ount until the morrow of the seventh Sabbath. You shall count [fifty] days. You shall bring a new grain-offering to YHWH from your homes, [a loaf of fine fl]ou[r], freshly baked with leaven. They are firstfruits to YHWH, wheat bread, twe[lve cakes, two] tenths of fine flour in each cake ... the tribes of Israel. They shall offer XIX ... their [grain-offerin]g and dr[ink-offering] according to the statute. The [priests] shall wave ... [wave-offering with the bread of] the firstfruits. They shall b[elong to] the priests and they shall eat them in the [inner] court[yard], [as a ne]w [grain-offering], the bread of the firstfruits. Then ... new bread from freshly ripened ears. [On this] da[y] there shall be [a holy gathering, an eter]nal [rule] for their generations. [They] shall [do] no work. It is the feast of Weeks and the feast of Firstfruits, an eterna[l] memorial. You [shall count] seven weeks from the day when you bring the new grain-offering to YHW [H], the bread of firstfruits. Seven full Sabbaths [shall elapse un]til you have counted fifty days to the morrow of the seventh Sabbath. [You] shall [bring] new wine for a drink-offering, four hins from all the tribes of Israel, one third of a hin for each tribe. They shall offer on this day with the wine twelve rams to YHWH; all the chiefs of the clans of Israel XX ... [r]ams and the corresponding grain-offering according to the statute: two [tenths of fine flour mixed with oil, one third of a h]in of oil for a ram; with this drink-offering ... seven yearling ram lambs and a he-[goat] ... assembly ... their [grainoffering and drink-offering] (shall be) according to the statute concerning young bulls and the ram ... to YHWH. At the quarter of the day, they shall offer ... [the r]ams and the drink-offering. They shall offer ... fourteen yearling ram lambs ... the burnt-offering. They shall prepare them ... and they shall burn their fat on the altar, [the fat covering the entrails] and the fat that is on them, and [the appendage of the liver with] the kidneys he shall remove and the fat on [them], and that which is on the loins and the fat tail close to the backbone. They shall b[urn all on the altar] together with the corresponding grain-offering and drinkoffering, an offering by fire, of soothing odou[r before YHWH]. They shall offer every grain-offering joined to a drink-offering according to [the statute]. They shall take a handful from [eve]ry grain-offering offered either with frankincense or dry, (this being) its [memorial portion], and burn it on the altar. They shall eat the remainder in the [in]n[er] courtyard. The priests shall e[a]t it unleavened. It shall not be eaten with leaven. It shall be ea[ten] on that day [before] sun[se]t. They shall salt all their offerings. You shall never allow the covenant of salt to fail. They shall offer to YHWH an offering from the rams and the lambs, the right thigh, the breast, [the cheeks, the stomac]h and the foreleg as far as the shoulder bone, and they shall wave them as a wave-offering. XXI [The priests'] portions [shall] be the thigh of the offering and the breast ... [the foreleg]s, the cheeks and the stomachs ... [as an eternal rule, from the children of Isra]el and the shoulder remaining of the foreleg [shall be for the Levites] ... an eternal rule for them and for their seed .. the princes of the Thousands ... [from] the rams and from [the lambs, one ram and one ram lamb (shall belong) to the priests; to the Levites], one [ra]m, one lamb; and to every [tribe, on]e [ram], one lamb for all the tri[bes], the [twe]lve tribes of Israel. They shall eat them [on that day, in the out]er [courtyard] before YHWH. ... [the priest]s shall drink there first and the Levites [second] ... the princes of the standards first ... [men of] renown. After them the whole people, from the great to the small, shall begin to drink the new wine. They [shall not e]a[t] any un[ri]pe grapes from the vines, for [on] this [da]y they shall expiate for the tirosh. The children of Israel shall rejoice before YHWH, an eternal [rule] for their generations wherever they dwell. They shall rejoice on [this] d[ay for they have begun] to pour out an intoxicating drink-offering, the new wine, on the altar of YHWH, year by year. [You sha]ll count from that day seven weeks, seven times (seven days), forty-nine days; there shall be seven full Sabbaths; until the morrow of the seventh Sabbath you shall count fifty days. You shall then offer new oil from the homes of [the tr]ibes of the ch[ildren of Is]rael, half a hin from a tribe, new beaten oil ... oil on the altar of the holocaust, firstfruits before YHWH. XXII ... [shall expi]ate with it for all the congregation before [YHWH] ... with this oil, half a hin ... [according to the st]atute, a holocaust, an offering by fire, of soothing [odour to YHWH] ... [With] this oil they shall light the lamps ... the princes of the Thousands with ... fourteen [yearling] m[ale lamb]s and the corresponding grain-offering and drink-offering ... [for the lambs and] the rams. The Levites shall slaughter ... [and] the priests, the sons of Aaron, [shall spri]nkle their blood [on the altar all around] ... [and] they shall burn their fat on the altar of the [holocaust] ... [and the corresponding grain-offering] and drink-offering, they shall burn over the fats ... [an offering by fire, of soothing odour to] YHWH. They shall take away fr[om] ... the right thigh and the breast ... the cheeks and the stomach shall be the priests' portion according to the statute concerning them. (They shall give) to the Levites the shoulder. Afterwards they shall bring them (the offerings) out to the children of Israel, and the children of Israel shall give the prie[st]s one ram, one lamb, and to the Levites, one ram, one lamb, and to each tribe, one ram, one lamb. They shall eat them on that day in the outer courtyard before YHWH, an eternal rule for their generations, year by year. Afterwards they shall eat from the olives and anoint themselves with the new oil, for on this day they shall expiate for [al]l [the o]il of the land before YHWH once yearly. They shall rejoice XXIII ... The High Priest shall offer the [holocaust of the Levites] first, and afterwards he shall send up in smoke the holocaust of the tribe of Judah, and w[hen he] is sending it up in smoke, they shall slaughter before him the he-goat first and he shall lift up its blood in a bowl to the altar and with his finger he shall pu[t some] of the blood to the four horns of the alta[r] of the holocaust and to the four corners of the altar ledge, and shall toss the blood towards the bas[e] of the altar ledge all around. He shall burn its fat on the altar, the fat covering the entrails and that over the entrails. The appendage of the liver with the kidneys he shall remove as well as the fat over them and on the loins. He shall send up in smoke all of them on the altar together with the corresponding grain-offering and drink-offering, an offering by fire of soothing odour to YHWH. And XXIV ... the flesh, of [soothing] odour; it shall be [an offering by fire to YHWH. Thus they must do to every] young bull, and to every ram and to [every lamb] and its limbs (?) shall remain apart. The corresponding [grain-offering] and drink-offering shall be on it, an [eternal] rule for your generations before YHWH. After this holocaust he shall offer the holocaust of the tribe of Judah separately. As he has done with the holocaust of the Levites, so shall he do with the holocaust of the children of Judah after the Levites. On the second day he shall first offer the holocaust of Benjamin and after it he shall offer the holocaust of the children of Joseph, Ephraim and Manasseh together. On the third day, he shall offer the holocaust of Reuben separately, and the holocaust of Simeon separately. On the fourth day he shall offer the holocaust of Issachar separately and the holocaust of Zebulun separately. On the fifth day he shall offer the holocaust of Gad separately and the holocaust of Asher separately. On the sixth day XXV [he shall offer the holocaust of Dan separately and the holocaust of Naphtali separately] ... In the [seventh] m[onth, on the first day of the month, you shall have] a sacred rest, a remembrance announced by a trumpet blast, a [holy] ga[thering. You shall offer a holocaust, an offering by fire, of soothing odour be]fore YHWH. You shall o[ffer on]e [young bull,] one ram, seve[n] ye[ar]ling [lamb]s [without blemish and one he-goat for a sinoffering, and] the corresponding grain-offering and drink-offering according to the statute concerning the[m, of soothing odour to YHWH, in addition to] the perpetual [holocaus]t [and the holo]caust of the new moon. Afterwards [you shall offer] this [holocaust] at the third part of the day, an eternal rule for your generation[s wherever you dwell.] You shall rejoice on this day. On it you shall do no work. A sacred rest shall this day be for you. The tenth of this month is the Day of Atonement. You shall mortify yourselves. For any person who does not mortify himself on this selfsame day shall be cut off from his people. You shall offer on it a holocaust to YHWH: one young bull, one ram, seven ram lambs, one he-goat for a sin-offering, in addition to the sin-offering of the atonement and the corresponding grain-offering and drink-offering according to the statute concerning the young bull, the ram, the lambs and the he-goat. For the sin-offering of the atonement you shall offer two rams for holocaust. The High Priest shall offer one for himself and his father's house XXVI ... [The High Prie]st [shall cast lots on the two goats,] o[ne] lot for YHWH and one for Azazel. He shall slaughter the goat [on] which [YHWH's lot has fallen and shall lift up] its blood in a golden bowl which is in [his ha]nd, [and do] with its blo[od as he has done with the blood of] his young bull and shall expiate with it for all the people of the assembly. He shall send up in smoke its fat and the corresponding grain-and drink-offering on the altar of the holocaust. Its flesh, skin and dung they shall burn beside his young bull. It is a sinoffering for the whole assembly. He shall expiate with it for all the people of the assembly and it shall be forgiven to them. He shall wash his hands and feet of the blood of the sin-offering and shall come to the living goat and shall confess over its head the iniquities of the children of Israel together with all their guilt, all their sins. He shall put them on the head of the goat and despatch it to Azazel in the desert by the hand of the man who is waiting ready. The goat shall bear all the iniquities of (the children of Israel). XXVII ... [and he shall expiate] for all the children of Israel and it shall be forgiven to them ... Afterwards he shall offer the young bull, the r[a]m, and [the lambs, according to] the [sta]tute relating to them, on the altar of the holocaust, and the [ho]locaust will be accepted for the children of Israel, an eternal rule for their generations. Once a year this day shall be for them a memorial. They shall do no work on it, for it shall be [to] them a Sabbath of sacred rest. Whoever shall do work on it or shall not mortify himself on it, shall be cut off from the midst of his people. A Sabbath of sacred rest, a holy gathering shall this day be for you. You shall sanctify it as a memorial wherever you dwell and you shall do no work. On the fifteenth day of this month XXVIII ... [the corresponding] grainoffering [and drink-offering, all on] the altar, an offering by fire, of s[oothing odour to YHWH. On] the second [day:] twelve young bulls, [two rams, four]teen [lambs] and one he-goat [for a sin-offerin]g [and the corresponding gr]ai[n-offering and drink-offering] according to the statute concerning the young bulls, the ram[s], the lambs [and] the hegoat; it is an offering by fire, of soothing odour to YHWH. On the third day eleven young bulls, two rams, fourteen lambs and one he-goat for a sin-offering and the corresponding grain-offering and drink-offering according to the statute concerning the young bulls, the rams, the lambs and the he-goat. On the fo[ur]th day ten young bulls, two rams, fourteen yearling ram lambs and one he-goat for a sin-offering and the corresponding grainoffering and drink-offering for the young bulls, XXIX [the rams, the lambs and the he-goat ... On the fifth day ... and the corresponding grain-offering] and drink-offer[ing] ... in the house on which I [shall cause] my name to rest ... holocausts, [each on its] day according to the law of this statute, always from the children of Israel in addition to their freewillofferings in regard to all that they offer, their drink-offerings and all their gifts that they shall bring to me in order to be acceptable. I shall accept them and they shall be my people and I shall be for them for ever. I will dwell with them for ever and ever and will sanctify my [sa]nctuary by my glory. Iwill cause my glory to rest on it until the day of creation on which I shall create my sanctuary, establishing it for myself for all time according to the covenant which I have made with Jacob in Bethel. XXX ... You shall make ... for stairs, a stair[case] ... in the house which you shall build ... You [shall make] a staircase north of the Temple, a square house, twenty cubits from one corner to the other alongside its four corners. Its distance from the wall of the Temple shall be seven cubits on the north-west. You shall make the width of its wall four cubits ... like the Temple and its inside from corner to corner twelv[e cubits.] (There shall be) a square column in its middle, in the centre; its width four cubits on each side around which the stairs wind ... XXXI In the upper chamber of [this] ho[use you shall make a ga]te opening to the roof of the Temple and a way (shall be) made through this gate towards the entrance ... of the Temple by which one can reach the upper chamber of the Temple. Overlay with gold [a]ll this stairhouse, its walls, its gates and its roof, from inside [and from] outside, its column and its stairs. [You] shall do everything as I tell you. You shall make a square house for the laver in the south-east, on all its sides, (each) twenty-one cubits; fifty cubits distant from the altar. The width of the wall shall be four cubits, and the height [t]wenty cubits ... Make gates for it on the east, on the north and on the west. The width of the gates shall be four cubits and the height seven XXXII ... You shall make in the wall of this house, on the inside, recesses, and in them ... one cubit (in) width and their height four cubits above the ground. They shall be overlaid with gold on which they shall place their clothes which they have worn on arrival. Above the house of the ... when they come to minister in the sanctuary. You shall make a trench around the laver beside its house and the trench shall go [from the house of] the laver to a cavity. It shall descend [rapid]ly to the ground where the water shall flow and disappear. It shall not be touched by any man for it is mingled with the blood of the holocaust. XXXIII They shall sanctify my people in the sacred vestments which ... You shall make a house east of the house of the [l]av[er] according to the measurement of [the house of the bas]in. Its wall shall be at a distance of seven cubits from the wall of the house of the laver. Its whole building and rafters shall be like (those of) the house of the laver. It shall have two gates on the north and the south, one opposite the other, according to the measurement of the gates of the house of the laver. Inside all the walls of this house shall have apertures, their width (and depth) two cubits each and their height four (?) with which the entrails and the feet are raised to the altar. When they have completed the sending up in smoke XXXIV ... They close the wheels and ... and tie the horns of the young bulls to the rings and ... by the rings. Afterwards they shall slaughter them and collect [the blood] in bowls and toss it around the altar base. They shall open the wheels and strip the skin of the young bulls from their flesh and cut them up into pieces, salt the pieces, wash the entrails and the legs, salt them and send them up in smoke on the fire which is on the altar, each young bull with its pieces beside it and the corresponding grain-offering of fine flour on it, the wine of the drink-offering beside it and some of it on it. The priests, the sons of Aaron, shall send everything up in smoke on the altar, an offering by fire, of soothing odour before YHWH. You shall make chains hanging from the rafters of the twelve columns XXXV ... whoever is not a priest shall die, and whoever ... [a prie]st who shall come ... and he is not clothed in the [holy] vest[ments in which] he was ordained, they too shall be put to death and shall not pro[fane the san]ctuary of their God, thus incurring the iniquity of mortal guilt. You shall sanctify the environs of the altar, the Temple, the laver and the colonnade and they shall be most holy for ever and ever. You shall make a place west of the Temple, a colonnade of pillars standing around for the sin-offerings and the guilt-offerings, divided from one another, the sin-offerings of the priests, the he-goats, and the sin-offerings of the people and their guilt-offerings. None of these shall be mingled one with another, for their places shall be divided from one another in order that the priests may not err concerning all the sinofferings of the people, and all the rams (?) of the guilt-offerings, (thus) incurring the sin of guilt. The birds for the altar: he shall prepare turtledoves XXXVI ... from the corner of ... [to the corne]r of the gat[e, one hundred and twenty cubits.] The gate (shall be) forty [cubits] wide. Each side shall be [according to this measurement. The wid]th of [its wa]ll shall be seven cubits, [and] its [height forty]-five [cubits to the raft]ers of [its] roof. The width of its ch[ambers] (shall be) twenty-six cubits from corner to corner. The gates of entrance and exit: the gate shall be fourteen cubits wide and [tw]enty-eight cubits high from the threshold to the lintel. The height of the rafters above the lintel shall be fourteen cubits. (The gate shall be) roofed with a panelling of cedar wood overlaid with pure gold. Its doors shall be overlaid with fine gold. From the corner of the gate to the second angle of the courtyard, (there shall be) one hundred and twenty cubits. Thus shall be the measurement of all these gates of the inner courtyard. The gates shall lead inside into the courtyard. XXXVII You shall make [in]side the court[yard] seats for the priests, and tables in front of the seats, in the inner colonnade by the outer wall of the courtyard, places made for the priests and their sacrifices, for the firstfruits and the tithes, for their peace-offering sacrifices which they shall sacrifice. The sacrifices of the peace-offerings of the children of Israel shall not be mingled with the sacrifices of the priests. In the four corners of the courtyard you shall make for them a place for cooking-stoves where they shall seethe their sacrifices [and] sinofferings. XXXVIII ... There they shall eat ... the bird, the turtle-dove and the young pigeons ... You shall make a second [co]urtyard aro[u]nd [the in]ner [courtyard], one hundred cubits wide, and four hundred and eighty cubits long on the east side, and thus shall be the width and length of all its sides: to the south, to the west and to the north. Its wall shall be [fo]ur cubits wide and twenty-eight cubits high. Chambers shall be made in the wall outside and between each chamber there shall be three-[and-a-half cubits] XXXIX ... that all the congregation of the children of Israel may bow down before me ... No woman shall come there, nor a child until the day that he has fulfilled the rule ... [and has paid for] himself [a ransom] to YHWH, half a shekel, an eternal rule, a memorial wherever they dwell. The shekel (consists of) twenty gerahs. When they shall collect from him the half-shekel... to me. Afterwards they shall enter from the age of twenty ... The na[mes of the g]ates of this [co]urtyard sha[ll b]e according to the nam[es of] the children of Is[ra]el: Simeon, Levi and Judah in the east; Reuben, Joseph and Benjamin in the south; Issachar, Zebulun and Gad in the west; Dan, Naphtali and Asher in the north. Between each gate the measurement (shall be): from the north-eastern corner to the gate of Simeon, ninetynine cubits, and the gate twenty-eight cubits. From this gate of Simeon to the gate of Levi, ninety-nine cubits, and the gate, twenty-eight cubits. From the gate of Levi to the gate of Judah XL ... You shall make a third courtyard ... to their daughters and to the strangers who [were] born ... [wi]de around the middle courtyard ... in length about one thousand six [hundred] cubits from one corner to the next. Each side shall be according to this measurement: on the east, the south, the west and the no[rt]h. The wall shall be seven cubits wide and forty-nine cubits high. Chambers shall be made between its gates along the foundation as far up as its 'crowns' (= crenellations: Yadin). There shall be three gates in the east, three in the south, three in the west and three in the north. The gates shall be fifty cubits wide and their height seventy cubits. Between one gate and another there shall be three hundred and sixty cubits. From the corner to the gate of Simeon, three hundred and sixty cubits. From the gate of Simeon to the gate of Levi, likewise. From the gate of Levi to the gate of Judah, likewise three [hundred and] sixty (cubits). XLI ... From the gate of Issachar [to the gate of Zebulun, three] hundred [and sixty] cubits. From the gate of Zebulun to the gate of Gad, three hundred and sixty cubits. From the ga[te of] Gad to the northern corner, three hundred and sixty cubits. From this corner to the gate of Dan: three hundred and sixty cubits. Thus from the gate of Dan to the gate of Naphtali, three hundred and sixty cubits. From the gate of Naphtali to the gate of Asher, three hundred and sixty cubits. From the gate of Asher to the eastern corner, three hundred and sixty cubits. The gates shall jut outwards from the wall of the courtyard seven cubits, and extend inwards from the wall to the courtyard thirty-six cubits. The entrance of the gate shall be fourteen cubits wide and twenty-eight cubits high up to the lintel. The rafters at the doorways (?) shall be of cedar wood and overlaid with gold. The doors shall be overlaid with pure gold. Between each gate inwards you shall make storehouses, XLII [rooms and colonnades.]

The room shall be ten cubits wide, twenty cubits long, and four[teen] cubits high ... with cedar wood. The wall shall be two cubits wide. On the outside there shall be storehouses. [The storehouse shall be ten cubits wide and] twenty cubits [long]. The wall shall be two cubits wide [and fourteen cubits high] up to the lintel. Its entrance shall be three cubits wide. [You shall make in this way] all the storehouses and the [corresponding] rooms. The colon[nade] ... shall be ten cubits [wi]de. Between each gate [you shall make eight]een storehouses and the corresponding eight[een] rooms ... You shall make a staircase next to the walls of the gates towards the colonnade. Winding stairs shall go up to the second and third colonnades and to the roof. You shall build storehouses and corresponding rooms and colonnades as on the ground floor. The second and the third (levels) shall follow the measurement of the lower one. On the roof of the third you shall make pillars roofed with rafters from one pillar to the next (providing) a place for tabernacles. The (pillars) shall be eight cubits high and the tabernacles shall be made on their (roof) each year at the feast of the Tabernacles for the elders of the congregation, for the princes, the heads of the fathers' houses of the children of Israel, the captains of the thousands, the captains of the hundreds, who will ascend and dwell there until the sacrificing of the holocaust on the festival which is the feast of the Tabernacles, each year. Between each gate there shall be XLIII ... on the days of the firstfruits of the corn, of the w[ine (tirosh) and the oil, and at the festival of the offering of] wood. On these days (the tithe) shall be eaten. They shall not put aside anything from it from one year to another. For they shall eat it in this manner. From the feast of the Firstfruits of the corn of wheat they shall eat the corn until the next year, until the feast of the Firstfruits, and (they shall drink) the wine from the day of the festival of Wine until the next year, until the day of the festival of the Wine, and (they shall eat) the oil from its festival, until the next year, until the festival, the day of offering the new oil on the altar. Whatever is left (to last beyond) their festivals shall be sanctified by being burnt with fire. It shall no longer be eaten for it is holy. Those who live within a distance of three days' walk from the sanctuary shall bring whatever they can bring. If they cannot carry it, they shall sell it for money and buy with it corn, wine, oil, cattle and sheep, and shall eat them on the days of the festivals. On working days they shall not eat from this in their weariness for it is holy. On the holy days it shall be eaten, but it shall not be eaten on working days. XLIV ... You shall allot [the rooms and the corresponding chambers. From the gate of Simeo]n to the gate of Judah shall be for the priests ... All that is to the right and to the left of the gate of Levi, you shall allo[t] to Aaron, your brother, one hundred and eight rooms and corresponding chambers and two tabernacles which are on the roof. (You shall allot) to the sons of Judah (the area) from the gate of Judah to the corner: fifty-four rooms and corresponding chambers and the tabernacle that is over them. (You shall allot) to the sons of Simeon (the area) from the gate of Simeon to the second corner: their rooms, the corresponding chambers and tabernacles. (You shall allot) to the sons of Reuben (the area) from the corner which is beside the sons of Judah to the gate of Reuben: fifty-two rooms and the corresponding chambers and tabernacles. (The area) from the gate of Reuben to the gate of Joseph (you shall allot) to the sons of Joseph, to Ephraim and Manasseh. (The area) from the gate of Joseph to the gate of Benjamin (you shall allot) to the sons of Kohath from the Levites. (The area) from the gate of Benjamin to the western corner (you shall allot) to the sons of Benjamin. (The area) from this corner to the gate of Issachar (you shall allot) to the sons of Issachar. (The area) from the gate (of Issachar) XLV... the second (= incoming) [priestly course] shall enter on the left ... and the first (= outgoing) shall leave on the right. They shall not mingle with one another nor their vessels. [Each] priestly course shall come to its place and they shall stay there. One shall arrive and the other leave on the eighth day. They shall clean the rooms, one after the other, when the first (priestly course) leaves. There shall be no mingling there. No man who has had a nocturnal emission shall enter the sanctuary at all until three days have elapsed. He shall wash his garments and bathe on the first day and on the third day he shall wash his garments and bathe, and after sunset he shall enter the sanctuary. They shall not enter my sanctuary in their impure uncleanness and render it unclean. No man who has had sexual intercourse with his wife shall enter anywhere into the city of the sanctuary where I cause my name to abide, for three days. No blind man shall enter it in all his days and shall not profane the city where I abide, for I, YHWH, abide amongst the children of Israel for ever and ever. Whoever is to purify himself of his flux shall count seven days for his purification. He shall wash his garments on the seventh day and bathe his whole body in running water. Afterwards he shall enter the city of the sanctuary. No one unclean through contact with a corpse shall enter there until he has purified himself. No leper nor any man smitten (in his body) shall enter there until he has purified himself and has offered ... XLVI ... [No] unclean bird shall fly over [my] sanctua[ry] ... the roofs of the gates ... the outer courtyard ... be in my sanctuary for ever and ever all the time that I [abide] among them. You shall make a terrace round about, outside the outer courtyard, fourteen cubits wide like the entrances of all the gates. You shall make twelve steps (leading) to it by which the children of Israel shall ascend there to enter my sanctuary. You shall make a one-hundred-cubits-wide ditch around the sanctuary which shall divide the holy sanctuary from the city so that no one can rush into my sanctuary and defile it. They shall sanctify my sanctuary and hold it in awe because I abide among them. You shall make for them latrines outside the city where they shall go out, north-west of the city. These shall be roofed houses with holes in them into which the filth shall go down. It shall be far enough not to be visible from the city, (at) three thousand cubits. You shall make three areas to the east of the city, divided from one another, where the lepers, those suffering from a flux and men who have had a (nocturnal) emission XLVII ... Their cities [shall be] pure ... for ever. The city which I will sanctify, causing my name and [my] sanctuar[y] to abide [in it], shall be holy and pure of all impurity with which they can become impure. Whatever is in it shall be pure. Whatever enters it shall be pure: wine, oil, all food and all moistened (food) shall be clean. No skin of clean animals slaughtered in their cities shall be brought there (to the city of the sanctuary). But in their cities they may use them for any work they need. But they shall not bring them to the city of my sanctuary, for the purity of the skin corresponds to that of the flesh. You shall not profane the city where I cause my name and my sanctuary to abide. For it is in the skins (of animals) slaughtered in the sanctuary that they shall bring their wine and oil and all their food to the city of my sanctuary. They shall not pollute my sanctuary with the skins of animals slaughtered in their country which are tainted (= unfit for the Temple). You cannot render any city among your cities as pure as my city, for the purity of the skin of the animal corresponds to the purity of its flesh. If you slaughter it in my sanctuary, it shall be pure for my sanctuary, but if you slaughter it in your cities, it shall be pure (only) for your cities. Whatever is pure for the sanctuary, shall be brought in skins (fit) for the sanctuary, and you shall not profane my sanctuary and my city where I abide with tainted skins. XLVIII ... [the cormorant, the stork, every ki]nd of [heron,] the hoop[oe and the bat] ... You may eat [the following] flying [insects]: every kind of great locust, every kind of long-headed locust, every kind of green locust, and every kind of desert locust. These are among the flying insects which you may eat: those which walk on four legs and have legs jointed above their feet to leap with them on the ground and wings to fly with. You shall not eat the carcass of any bird or beast but may sell it to a foreigner. You shall not eat any abominable thing, for you are a holy people to YHWH, your God. You are the sons of YHWH, your God. You shall not gash yourselves or shave your forelocks in mourning for the dead, nor shall you tattoo yourselves, for you are a holy people to YHWH, your God. You shall not profane your land. You shall not do as the nations do; they bury their dead everywhere, they bury them even in their houses. Rather you shall set apart areas in the midst of your land where you shall bury your dead. Between four cities you shall designate an area for burial. In every city you shall set aside areas for those stricken with leprosy, with plague and with scab, who shall not enter your cities and profane them, and also for those who suffer from a flux; and for menstruating women, and women after childbirth, so that they may not cause defilement in their midst by their impure uncleanness. The leper suffering from chronic leprosy or scab, who has been pronounced unclean by the priest XLIX ... with cedar wood, hyssop and ... your cities with the plague of leprosy and they shall be unclean. If a man dies in your cities, the house in which the dead man has died shall be unclean for seven days. Whatever is in the house and whoever enters the house shall be unclean for seven days. Any food on which water has been poured shall be unclean, anything moistened shall be unclean. Earthenware vessels shall be unclean and whatever they contain shall be unclean for every clean man. The open (vessels) shall be unclean for every Israelite (with) whatever is moistened in them. On the day when the body is removed from there, they shall cleanse the house of all pollution of oil, wine and water moisture. They shall rub its (the house's) floor, walls and doors and shall wash with water the bolts, doorposts, thresholds and lintels. On the day when the body is removed from there, they shall purify the house and all its utensils, hand-mills and mortars, all utensils of wood, iron and bronze and all utensils capable of purification. Clothes, sacks and skins shall be washed. As for the people, whoever has been in the house or has entered the house shall

196

bathe in water and shall wash his clothes on the first day. On the third day they shall sprinkle purifying water on them and shall bathe. They shall wash their garments and all the utensils in the house. On the seventh day they shall sprinkle (them) a second time. They shall bathe, wash their clothes and utensils and shall be clean by the evening of (the impurity contracted) from the dead so as to (be fit to) touch their pure things. As for a man who has not been rendered unclean on account of L ... they have been unclean. No longer ... until they have sprinkled (them) the second [time] on the seventh day and shall be clean by the evening at sunset. Whoever touches the bone of a dead person in the fields, or one slain by the sword, or a dead body or the blood of a dead person, or a tomb, he shall purify himself according to the rule of this statute. But if he does not purify himself according to the statute of this law, he is unclean, his uncleanness being still in him. Whoever touches him must wash his clothes, bathe and he shall be clean by the evening. If a woman is with child and it dies in her womb, as long as it is dead in her, she shall be unclean like a tomb. Any house that she enters shall be unclean with all its utensils for seven days. Whoever touches it shall be unclean till the evening. If anyone enters the house with her, he shall be unclean for seven days. He shall wash his clothes and bathe in water on the first (day). On the third day he shall sprinkle and wash his clothes and bathe. On the seventh day he shall sprinkle a second time and wash his clothes and bathe. At sunset he shall be clean. As for all the utensils, clothes, skins and all the materials made of goat's hair, you shall deal with them according to the statute of this law. All earthenware vessels shall be broken for they are unclean and can no more be purified ever. All creatures that teem on the ground you shall proclaim unclean: the weasel, the mouse, every kind of lizard, the wall gecko, the sand gecko, the great lizard and the chameleon. Whoever touches them dead LI ... [and whatever com]es out of the[m] ... [shall be] unclean [to you.] You shall [not] render yourselves unclean by th[em. Whoever touches them] dead shall be unclean un[til the] evening. He shall wash his clothes and bathe [in water and at] sun[set] he shall be clean. Whoever carries any of their bones, their carcass, skin, flesh or claw shall wash his clothes and bathe in water. After sunset he shall be clean. You shall forewarn the children of Israel about all the impurities. They shall not render themselves unclean by those of which I tell you on this mountain and they shall not be unclean. For I, YHWH, abide among the children of Israel. You shall sanctify them and they shall be holy. They shall not render themselves abominable by anything that I have separated for them as unclean and they shall be holy. You shall establish judges and officers in all your towns and they shall judge the people with just judgement. They shall not be partial in (their) judgement. They shall not accept bribes, nor shall they twist judgement, for the bribe twists judgement, overturns the works of justice, blinds the eyes of the wise, produces great guilt, and profanes the house by the iniquity of sin. Justice and justice alone shall you pursue that you may live and come to inherit the land that I give you to inherit for all days. The man who accepts bribes and twists just judgement shall be put to death. You shall not be afraid to execute him. You shall not do in your land as the nations do. Everywhere they sacrifice, plant sacred trees, erect sacred pillars and set up carved stones to bow down before them and build for them LII ... You shall not plant [any tree as a sacred tree beside my altar to be made by you.] You shall not erect a sacred pillar [that is hateful to me.] You shall not make anywhere in your land a carved stone to bow down before it. You shall not sacrifice to me any cattle or sheep with a grave blemish, for they are abominable to me. You shall not sacrifice to me any cattle or sheep or goat that is pregnant, for this would be an abomination to me. You shall not slaughter a cow or a ewe and its young on the same day, neither shall you kill a mother with her young. Of all the firstlings born to your cattle or sheep, you shall sanctify for me the male animals. You shall not use the firstling of your cattle for work, nor shall you shear the firstling of your small cattle. You shall eat it before me every year in the place that I shall choose. Should it be blemished, being lame or blind or (afflicted with) any grave blemish, you shall not sacrifice it to me. It is within your towns that you shall eat it. The unclean and the clean among you together (may eat it) like a gazelle or a deer. It is the blood alone that you shall not eat. You shall spill it on the ground like water and cover it with dust. You shall not muzzle an ox while it is threshing. You shall not plough with an ox and an ass (harnessed) together. You shall not slaughter clean cattle or sheep or goat in any of your towns, within a distance of three days' journey from my sanctuary. It is rather in my sanctuary that you shall slaughter it, making of it a holocaust or peace-offering. You shall eat and rejoice before me in the place on which I choose to set my name. Every clean animal with a blemish, you shall eat it within your towns, away from my sanctuary at a distance of thirty stadia. You shall not slaughter it close to my sanctuary for its flesh is tainted. You shall not eat in my city, which I sanctify by placing my name in it, the flesh of cattle, sheep or goat which has not entered my sanctuary. They shall sacrifice it there, toss its blood to the base of the altar of holocaust and shall burn its fat. LIII [When I extend your frontiers as I have told you, and if the place where I have chosen to set my name is too distan]t, and you say, 'I will eat meat', because you [l]ong for it, [whatever you desire,] you may eat, [and you may slau]gh[ter] any of your small cattle or cattle which I give you according to my blessing. You may eat it within your towns, the clean and the unclean together, like gazelle or deer (meat). But you shall firmly abstain from eating the blood. You shall spill it on the ground like water and cover it with dust. For the blood is the life and you shall not eat the life with the flesh so that it may be well with you and with your sons after you for ever. You shall do that which is correct and good before me, for I am YHWH, your God. But all your devoted gifts and votive donations you shall bring when you come to the place where I cause my name to abide, and you shall sacrifice (them) there before me as you have devoted and vowed them with your mouth. When you make a vow, you shall not tarry in fulfilling it, for surely I will require it of you and you shall become guilty of a sin. You shall keep the word uttered by your lips, for your mouth has vowed freely to perform your vow. When a man makes a vow to me or swears an oath to take upon himself a binding obligation, he must not break his word. Whatever has been uttered by his mouth, he shall do it. When a woman makes a vow to me, or takes upon herself a binding obligation by means of an oath in her father's house, in her youth, if her father hears of her vow or the binding obligation which she has taken upon herself and remains silent, all her vows shall stand, and her binding obligation which she has taken upon herself shall stand. If, however, her father definitely forbids her on the day that he hears of it, none of her vows or binding obligations which she has taken upon herself shall stand, and I will absolve her because (her father) has forbidden her LIV [when he] h[eard of them. But if he annuls them after] the da[y that he has] hea[rd of them, he shall bear] her guilt: [her] fa[ther has annulled them. Any vow] or binding oath (made by a woman) [to mortify herself,] her husband may confi[rm it] or annul it on the day that he hears of it, and I will absolve her. But any vow of a widow or a divorced woman, whatever she has taken upon herself shall stand in conformity with all that her mouth has uttered. Everything that I command you today, see to it that it is kept. You shall not add to it, nor detract from it. If a prophet or a dreamer appears among you and presents you with a sign or a portent, even if the sign or the portent comes true, when he says, 'Let us go and worship other gods whom you have not known!', do not listen to the words of that prophet or that dreamer, for I am testing you to discover whether you love YHWH, the God of your fathers, with all your heart and soul. It is YHWH, your God, that you must follow and serve, and it is him that you must fear and his voice that you must obey, and you must hold fast to him. That prophet or dreamer shall be put to death for he has preached rebellion against YHWH, your God, who brought you out of the land of Egypt and redeemed you from the house of bondage, to lead you astray from the path that I have commanded you to follow. You shall rid yourself of this evil. If your brother, the son of your father or the son of your mother, or your son, or your daughter, or the wife of your bosom, or your friend who is like your own self, (seeks to) entice you secretly, saying, 'Let us go and worship other gods whom you have not known', neither you, LV [nor] your [fa]thers, some of the gods [of the peoples that are round about you, whether near you or far off from you], from the one end of the earth to [the other, you shall not yield to him or listen to him, nor shall your eye pity] him, nor shall you spare [him, nor shall you conceal him; but you shall kill him; your hand shall be first against him to put him to death, and afterwards the hand of all the people. You shall stone him to death with stones because he sought to] draw you away [from me who brought you out of the land of Egypt, out of the house of bondage. And all Israel shall hear, and fear, and never again do such an evil thing] among you. If in on[e of your cities in which I] give you to dw[ell] you hear this said: 'Men, [s]ons of [Beli]al have arisen in your midst and have led astray all the inhabitants of their city saying, "Let us go and worship gods whom you have not known!",' you shall inquire, search and investigate carefully. If the matter is proven true that such an abomination has been done in Israel, you shall surely put all the inhabitants of that city to the sword. You shall place it and all who are in it under the ban, and you shall put the beasts to the sword. You shall assemble all the booty in (the city) square and shall burn it with fire, the city and all the booty, as a whole-offering to YHWH, your God. It shall be a ruin for ever and shall never be rebuilt. Nothing from that which has been placed under the ban shall cleave to your hand so that I may turn from my hot anger and show you compassion. I will be compassionate to you and multiply you as I told your fathers, provided that you obey my voice, keeping all my commandments that I command you today, to do that which is correct and good before YHWH, your God. If among you, in one of your towns that I give you, there is found a man or a woman who does that which is wrong in my eyes by transgressing my covenant, and goes and worships other gods, and bows down before them, or before the sun or the moon, or all the host of heaven, if you are told about it, and you hear about this matter, you shall search and investigate it carefully. If the matter is proven true that such an abomination has been done in Israel, you shall lead out that man or that woman and stone him (to death) with stones. LVI... [You shall go to the Levitical priests o]r to the [j]u[dges then in office]; you shall seek their guidance and [they] shall pro[nounce on] the matter for which [you have sought their guidance, and they shall procl]aim the(ir) judgement to you. You shall act in conformity with the law that they proclaim to you and the saying that they declare to you from the book of the Law. They shall issue to you a proclamation in truth from the place where I choose to cause my name to abide. Be careful to do all that they teach you and act in conformity with the decision that they communicate to you. Do not stray from the law which they proclaim to you to the right or to the left. The man who does not listen but acts arrogantly without obeying the priest who is posted there to minister before me, or the judge, that man shall die. You shall rid Israel of evil. All the people shall hear of it and shall be awe-stricken, and none shall ever again be arrogant in Israel. When you enter the land which I give you, take possession of it, dwell in it and say, 'I will appoint a king over me as do all the nations around me!', you may surely appoint over you the king whom I will choose. It is from among your brothers that you shall appoint a king over you. You shall not appoint over you a foreigner who is not your brother. He (the king) shall definitely not acquire many horses, neither shall he lead the people back to Egypt for war to acquire many horses and much silver and gold, for I told you, 'You shall never again go back that way'. He shall not acquire many wives that they may not turn his heart away from me. He shall not acquire very much silver and gold. When he sits on the throne of his kingdom, they shall write for him this law from the book which is before the priests. LVII This is the law [that they shall write for him] ... [They shall count,] on the day that they appoint hi[m] king, the sons of Israel from the age of twenty to sixty years according to their standard (units). He shall install at their head captains of thousands, captains of hundreds, captains of fifties and captains of tens in all their cities. He shall select from among them one thousand by tribe to be with him: twelve thousand warriors who shall not leave him alone to be captured by the nations. All the selected men whom he has selected shall be men of truth, God-fearers, haters of unjust gain and mighty warriors. They shall be with him always, day and night. They shall guard him from anything sinful, and from any foreign nation in order not to be captured by them. The twelve princes of his people shall be with him, and twelve from among the priests, and from among the Levites twelve. They shall sit together with him to (proclaim) judgement and the law so that his heart shall not be lifted above them, and he shall do nothing without them concerning any affair. He shall not marry as wife any daughter of the nations, but shall take a wife for himself from his father's house, from his father's family. He shall not take another wife in addition to her, for she alone shall be with him all the time of her life. But if she dies, he may marry another from his father's house, from his family. He shall not twist judgement; he shall take no bribe to twist a just judgement and shall not covet a field or a vineyard, any riches or house, or anything desirable in Israel. He shall (not) rob LVIII ... When the king hears of any nation or people intent on plundering whatever belongs to Israel, he shall send for the captains of thousands and the captains of hundreds posted in the cities of Israel. They shall send with him (the captain) one tenth of the people to go with him (the king) to war against their enemies, and they shall go with him. But if a large force enters the land of Israel, they shall send with him one fifth of the warriors. If a king with chariots and horses and a large force (comes), they shall send with him one third of the warriors, and the two (remaining) divisions shall guard their city and their boundaries so that no marauders invade their land. If the war presses him (the king) hard, they shall send to him half of the people, the men of the army, but the (other) half of the people shall not be severed from their cities. If they triumph over their enemies, smash them, put them to the sword and carry away their booty, they shall give the king his tithe of this, the priests one thousandth and the Levites one hundredth from everything. They shall halve the rest between the combatants and their brothers whom they have left in their cities. If he (the king) goes to war against his enemies, one fifth of the people shall go with him, the warriors, all the mighty men of valour. They shall avoid everything unclean, everything shameful, every iniquity and guilt. He shall not go until he has presented himself before the High Priest who shall inquire on his behalf for a decision by the Urim and Tummim. It is at his word that he shall go and at his word that he shall come, he and all the children of Israel who are with him. He shall not go following his heart's counsel until he (the High Priest) has inquired for a decision by the Urim and Tummim. He shall (then) succeed in all his ways on which he has set out according to the decision which LIX ... and they shall disperse them in many lands and they shall become a h[orror], a byword, a mockery. With a heavy yoke and in extreme want, they shall there serve gods made by human hands, of wood and stone, silver and gold. During this time their cities shall become a devastation, a laughing-stock and a wasteland, and their enemies shall devastate them. They shall sigh in the lands of their enemies and scream because of the heavy yoke. They shall cry out but I will not listen; they shall scream but I will not answer them because of their evil doings. I will hide my face from them and they shall become food, plunder and prey. None shall save them because of their wickedness, because they have broken my covenant and their soul has loathed my law until they have incurred every guilt. Afterwards they will return to me with all their heart and all their soul, in conformity with all the words of this law, and I will save them from the hand of their enemies and redeem them from the hand of those who hate them, and I will bring them to the land of their fathers. I will redeem them, and increase them and exult over them. I will be their God and they shall be my people. The king whose heart and eyes have gone astray from my commandments shall never have one to sit on the throne of his fathers, for I will cut off his posterity for ever so that it shall no more rule over Israel. But if he walk after my rules and keep my commandments and do that which is correct and good before me, no heir to the throne of the kingdom of Israel shall be cut off from among his sons for ever. I will be with him and will save him from the hand of those who hate him and from the hand of those who seek his life. I will place all his enemies before him and he shall rule over them according to his pleasure and they shall not rule over him. I will set him on an upward, not on a downward, course, to be the head and not the tail, that the days of his kingdom may be lengthened greatly for him and his sons after him. LX ... and all their wave-offerings. All their firstling male [bea]sts and all ... of their beasts and all their holy gifts which they shall sanctify to me together with all their holy gifts of praise and a proportion of their offering of birds, wild animals and fish, one thousandth of their catch, and all that they shall devote, and the proportion of the booty and the plunder. To the Levites shall belong the tithe of the corn, the wine and the oil that they have sanctified to me first; the shoulder from those who slaughter a sacrifice and a proportion of the booty, the plunder and the catch of birds, wild animals and fish, one hundredth; the tithe from the young pigeons and from the honey one fiftieth. To the priests shall belong one hundredth of the young pigeons, for I have chosen them from all your tribes to attend on me and minister (before me) and bless my name, he and his sons always. If a Levite come from any town anywhere in Israel where he sojourns to the place where I will choose to cause my name to abide, (if he come) with an eager soul, he may minister like his brethren the Levites who attend on me there. He shall have the same share of food with them, besides the inheritance from his father's family. When you enter the land which I give you, do not learn to practise the abominations of those nations. There shall be found among you none who makes his son or daughter pass through fire, nor an augur or a soothsayer, a diviner or a sorcerer, one who casts spells or a medium, or wizards or necromancers. For they are an abomination before me, all who practise such things, and it is because of these abominations that I drive them out before you. You shall be perfect towards YHWH, your God. For these nations that LXI ... to ut[ter a word] in [my] n[ame which I have n]ot comman[ded him to] utter, or wh[o speaks in the name of oth]er go[ds], that prophet shall be put to death. If you say in your heart, 'How shall we know the word which YHWH has not uttered?', when the word uttered by the prophet in the name of YHWH is not fulfilled and does not come true, that is not a word that I have uttered. The prophet has spoken arrogantly; do not fear him. A single witness may not come forward against a man in the matter of any iniquity or sin which he has committed. It is on the evidence of two witnesses or three witnesses that a case can be established. If a malicious witness comes forward against a man to testify against him in a case of a crime, both disputants shall stand before me and before the priests and the Levites and before the judges then in office, and the judges shall inquire, and if the witness is a false witness who has testified falsely against his brother, you shall do to him as he proposed to do to his brother. You shall rid yourselves of evil. The rest shall hear of it and shall be awe-stricken and never again shall such a thing be done in your midst. You shall have no mercy on him: life for life, eye for eye, tooth for tooth, hand for hand, foot for foot. When you go to war against your enemies, and you see horses and chariots and an army greater than yours, be not afraid of them, for I am with you who brought you out of the land of Egypt. When you approach the battle, the priest shall come forward to speak to the army and say to them, 'Hear, Israel, you approach ... ' ... LXII [and another man shall use its fruit. If any man has betrothed a woman but has not yet married her, he shall return] home. Otherwise he may die in the war and another man may take her. [The] of[ficers shall continue] to address the army and say, 'If any man is afraid and has lost heart, he shall go and return. Otherwise he may render his kinsmen as faint-hearted as himself.' When the judges have finished addressing the army, they shall appoint army captains at the head of the people. When you approach a city to fight it, (first) offer it peace. If it seeks peace and opens (its gates) to you, then all the people found in it shall become your forced labourers and shall serve you. If it does not make peace with you, but is ready to fight a war against you, you shall besiege it and I will deliver it into your hands. You shall put all its males to the sword, but the women, the children, the beasts and all that is in the city, all its booty, you may take as spoil for yourselves. You may enjoy the use of the booty of your enemies which I give you. Thus shall you treat the very distant cities, those which are not among the cities of these nations. But in the cities of the peoples which I give you as an inheritance, you shall not leave alive any creature. Indeed you shall utterly exterminate the Hittites, the Amorites, the Canaanites, the Hivites, the Jebusites, the Girgashites and the Perizzites as I have commanded you, that they may not teach you to practise all the abominations that they have performed to their gods. LXIII ... [a heifer with which] he has not worked, which [has not drawn the yoke. The elders of] that city [shall bring down] the heifer to a ravine with an ever-flowing stream which has never been sown or cultivated, and there they shall break its neck. The priests, the sons of Levi, shall come forward, for I have chosen them to minister before me and bless my name, and every dispute and every assault shall be decided by their word. All the elders of the city nearest to the body of the murdered man shall wash their hands over the head of the heifer whose neck has been broken in the ravine. They shall declare, 'Our hands did not shed this blood, nor did our eyes see it happen. Accept expiation for thy people Israel whom thou hast redeemed, O YHWH, and do not permit the guilt of innocent blood to rest among thy people, Israel. Let this blood be expiated for them.' You shall rid Israel (of the guilt) of innocent blood, and you shall do that which is correct and good before YHWH, your God. When you go to war against your enemies, and I deliver them into your hands, and you capture some of them, if you see among the captives a pretty woman and desire her, you may take her to be your wife. You shall bring her to your house, you shall shave her head, and cut her nails. You shall discard the clothes of her captivity and she shall dwell in your house, and bewail her father and mother for a full month. Afterwards you may go to her, consummate the marriage with her and she will be your wife. But she shall not touch whatever is pure for you for seven years, neither shall she eat of the sacrifice of peace-offering until seven years have elapsed. Afterwards she may eat. LXIV ... [the firstfruits of his virility; he has the right of the first-born.] If a man has a disobedient and rebellious son who refuses to listen to his father and mother, nor listens to them when they chastise him, his father and mother shall take hold of him and bring him to the elders of his city, to the gate of his place. They shall say to the elders of his town, 'This son of ours is disobedient and rebellious; he does not listen to us; he is a glutton and a drunkard.' All the men of his city shall stone him with stones and he shall die, and you shall rid yourselves of evil. All the children of Israel shall hear of it and be awe-stricken. If a man slanders his people and delivers his people to a foreign nation and does evil to his people, you shall hang him on a tree and he shall die. On the testimony of two witnesses and on the testimony of three witnesses he shall be put to death and they shall hang him on the tree. If a man is guilty of a capital crime and flees (abroad) to the nations, and curses his people, the children of Israel, you shall hang him also on the tree, and he shall die. But his body shall not stay overnight on the tree. Indeed you shall bury him on the same day. For he who is hanged on the tree is accursed of God and men. You shall not pollute the ground which I give you to inherit. If you see your kinsman's ox or sheep or donkey straying, do not neglect them; you shall indeed return them to your kinsman. If your kinsman does not live near you, and you do not know who he is, you shall bring the animal to your house and it shall be with you until he claims (it). LXV ... [Wh]en a bird's nest happens to lie before you by the roadside, on any tree or on the ground, with fledglings or eggs, and the hen is sitting on the fledglings or the eggs, you shall not take the hen with the young. You shall surely let the hen escape and take only the young so that it may be well with you and your days shall be prolonged. When you build a new house, you shall construct a parapet on the roof so that you do not bring blood-guilt on your house if anyone should fall from it. When a man takes a wife, has sexual intercourse with her and takes a dislike to her, and brings a baseless charge against her, ruining her reputation, and says, 'I have taken this woman, approached her, and did not find the proof of virginity in her', the father or the mother of the girl shall take the girl's proof of virginity and bring it to the elders at the gate. The girl's father shall say to the elders, 'I gave my daughter to be this man's wife; he has taken a dislike to her and has brought a baseless charge against her saying, "I have not found the proof of virginity in your daughter." Here is the proof of my daughter's virginity.' They shall spread out the garment before the elders of that city. The elders of that city shall take that man and chastise him. They shall fine him one hundred pieces of silver which they shall give to the father of the girl, because he (the husband) has tried to ruin the reputation of an Israelite virgin. He shall not LXVI ... [When a virgin betrothed to a man is found by another man in the city and he lies with her, they shall bring both of them to the gate] of that city and stone them with stones and they shall be put to death: the girl because she has not shouted (for help, although she was) in the city, and the man because he has dishonoured his neighbour's wife. You shall rid yourselves of evil. If the man has found the woman in the fields in a distant place hidden from the city, and raped her, only he who has lain with her shall be put to death. To the girl they shall do nothing since she has committed no crime worthy of death. For this affair is like that of a man who attacks his neighbour and murders him. For it was in the fields that he found her and the betrothed girl shouted (for help), but none came to her rescue. When a man seduces a virgin who is not betrothed, but is suitable to him according to the rule, and lies with her, and he is found out, he who has lain with her shall give the girl's father fifty pieces of silver and she shall be his wife. Because he has dishonoured her, he may not divorce her all his days. A man shall not take his father's wife and shall not lift his father's skirt. A man shall not take the wife of his brother and shall not lift the skirt of his brother, the son of his father or the son of his mother, for this is unclean. A man shall not take his sister, the daughter of his father or the daughter of his mother, for this is abominable. A man shall not take his father's sister or his mother's sister, for this is immoral. A man shall not take the daughter of his brother or the daughter of his sister for this is abominable. (A man) shall not take

Revelation 12

Introduction

Revelation 12 stands as one of the most compelling, evocative, and enigmatic chapters within the Book of Revelation, vividly capturing the spiritual imagination of countless readers, interpreters, and theologians throughout history. Rich with symbolic imagery, cosmic drama, and profound theological implications, this chapter portrays a celestial battle between cosmic forces, characterized by the appearance of a majestic woman clothed with the sun, the birth of a child destined to rule the nations, and the ferocious opposition of a great red dragon. Its dramatic depiction of spiritual warfare, divine intervention, persecution, protection, and ultimate victory embodies profound insights into the nature of good and evil, spiritual struggle, divine sovereignty, and human hope amid adversity.

The Book of Revelation itself—also known as the Apocalypse of John—holds a unique and authoritative place within Christian scripture, offering a visionary account of the cosmic struggle between divine and demonic forces, the unfolding of eschatological events, and the final triumph of God over evil. Traditionally attributed to the Apostle John and composed toward the end of the first century CE, Revelation is characterized by its vivid apocalyptic imagery, symbolic language, and prophetic visions, reflecting early Christian experiences of persecution, hope for divine intervention, and anticipation of final redemption. Within this broader literary and theological context, Revelation 12 occupies a pivotal narrative position, presenting symbolic archetypes and visionary themes foundational to the entire book's theological message.

The purpose of this introduction is to provide readers with a thorough historical, literary, and theological framework for engaging deeply with Revelation 12. By exploring its historical origins, analyzing its symbolic imagery and theological themes, and examining its enduring significance within Christian theology and spirituality, readers will gain invaluable insights into this remarkable biblical chapter. Despite its complex symbolism and interpretive challenges, Revelation 12 offers profound reflections on spiritual warfare, divine protection, human suffering, and ultimate victory, inviting readers to reflect deeply on humanity's enduring struggles, divine sovereignty, and eternal hope.

Historical and Literary Context

To fully appreciate Revelation 12, it is essential first to understand the historical and literary context in which it emerged. The Book of Revelation, written during the late first century CE, arose within a context of profound turmoil, persecution, and uncertainty for early Christian communities under Roman rule. During this period, Christians faced intense political, social, and religious pressures, including persecution, marginalization, and martyrdom. Revelation thus served as a powerful message of hope, encouragement, and spiritual perseverance, emphasizing divine sovereignty, ultimate justice, and eternal reward despite present hardships.

Attributed traditionally to the Apostle John during his exile on the island of Patmos, Revelation combines prophetic, apocalyptic, and epistolary literary elements, weaving together visionary experiences, symbolic narratives, and exhortative messages to early Christian communities. Its distinctive apocalyptic literary style—characterized by vivid symbolic imagery, celestial visions, angelic mediators, eschatological prophecies, and cosmic dualism—reflects widespread literary conventions within Jewish apocalyptic traditions, while simultaneously transforming and adapting these traditions to reflect Christian theological convictions and spiritual experiences.

Within this broader literary context, Revelation 12 functions as a central narrative transition, introducing cosmic themes, symbolic archetypes, and visionary motifs foundational to subsequent chapters. Its depiction of celestial conflict, spiritual warfare, persecution, and divine intervention provides readers with essential symbolic frameworks for interpreting Revelation's broader eschatological vision. Moreover, Revelation 12 resonates deeply with earlier biblical traditions and prophetic literature, echoing symbolic imagery, theological themes, and narrative motifs from the Hebrew Bible—particularly texts such as Genesis, Isaiah, Daniel, and Psalms—while simultaneously adapting and transforming these traditions within distinctly Christian theological perspectives.

Historically, Revelation 12 has been interpreted through various theological, historical, and symbolic lenses, reflecting diverse interpretive traditions within Christian theology. Early Christian interpreters often understood its symbolic figures as representations of the church's struggles against persecution, particularly under oppressive Roman rulers. Medieval and Reformation-era interpretations frequently emphasized allegorical, spiritual, or ecclesiological readings, identifying the woman as symbolic of the church, Mary, or both, and the dragon as emblematic of Satan, spiritual evil, or oppressive political authorities. Contemporary scholarly interpretations continue to explore historical, theological, and symbolic dimensions, emphasizing the chapter's profound insights into spiritual conflict, persecution, divine protection, and ultimate eschatological hope.

Symbolic Imagery and Theological Themes

Central to the enduring fascination, interpretive complexity, and theological richness of Revelation 12 are its vivid symbolic imagery and profound theological themes. The chapter opens dramatically with the appearance of a celestial woman clothed with the sun, standing upon the moon, and crowned with twelve stars—a majestic figure symbolizing divine favor, spiritual authority, and covenantal fulfillment. Interpretations of the woman's identity vary widely, including references to Israel, the church, Mary, or combinations thereof, underscoring the chapter's symbolic depth and interpretive complexity.

Significantly, the chapter's imagery echoes earlier biblical traditions, particularly Joseph's dream (Genesis 37), Israel's prophetic imagery (Isaiah, Micah), and celestial symbolism common in apocalyptic literature. The woman's pregnancy and the impending birth of a child who will rule the nations with an iron rod further evoke Messianic expectations, drawing upon Psalm 2 and prophetic traditions concerning divine kingship and eschatological redemption. Thus, the symbolic figure of the woman serves as a multifaceted archetype, embodying historical Israel's covenantal identity, the church's spiritual authority, Mary's role in salvation history, and eschatological hopes for divine intervention and redemption.

Opposing the woman and her child appears a great red dragon, vividly described as having seven heads, ten horns, and seven crowns, symbolizing power, authority, and destructive intent. The dragon explicitly represents Satan—the ancient adversary, deceiver, and accuser—seeking to devour the child and persecute the woman and her offspring. The ensuing celestial conflict vividly portrays spiritual warfare, divine intervention, and cosmic struggle between divine and demonic forces. Through symbolic language, the chapter emphasizes the pervasive reality of spiritual conflict, underscoring the destructive forces of evil, persecution, deception, and oppression confronting God's people.

Amid this cosmic battle, divine protection, spiritual preservation, and ultimate victory emerge as central theological themes. The woman and her child experience divine intervention and protection, including the child's ascension and the woman's refuge in the wilderness, symbolizing divine provision, spiritual safety, and eschatological hope despite persecution. Moreover, the angelic conflict and the dragon's expulsion from heaven underscore divine authority, cosmic justice, and ultimate victory over evil, emphasizing the triumph of divine sovereignty, redemption, and justice despite temporary hardships, persecution, or suffering.

Thus, Revelation 12 profoundly explores theological themes of spiritual warfare, divine sovereignty, human persecution, divine protection, and eschatological victory, offering readers powerful symbolic narratives emphasizing spiritual perseverance, moral resilience, and eternal hope amid earthly trials. Its symbolic imagery vividly illustrates cosmic realities, spiritual struggles, and ultimate divine triumph, inviting readers into profound theological reflection concerning humanity's perennial conflicts, spiritual aspirations, and divine assurances.

Enduring Significance and Contemporary Relevance

The historical and theological significance of Revelation 12 extends far beyond its original historical context, profoundly influencing Christian theology, spirituality, and cultural imagination throughout history. Historically, its vivid symbolic imagery, dramatic narrative, and profound theological themes significantly impacted early Christian eschatology, medieval spiritual traditions, Reformation-era theology, and contemporary Christian thought, shaping diverse theological interpretations, liturgical practices, devotional reflections, and artistic expressions.

The figure of the celestial woman profoundly influenced Marian devotion within medieval Christianity, shaping theological reflections concerning Mary's spiritual role, maternal symbolism, and eschatological significance. Additionally, its vivid portrayals of spiritual warfare, persecution, and divine victory profoundly shaped Christian understandings of evil, divine sovereignty, and spiritual resilience amid historical persecutions, societal conflicts, and theological controversies, offering communities powerful symbolic resources for spiritual reflection, moral courage, and theological reassurance.

Today, Revelation 12 continues to captivate scholarly interest and spiritual reflection, offering contemporary readers profound insights into spiritual warfare, divine-human interactions, persecution, and eschatological hope. Scholarly explorations increasingly appreciate its historical context, symbolic richness, theological depth, and contemporary relevance, highlighting its enduring importance for understanding Christian theology, spirituality, and ethical reflection amid contemporary cultural, political, and spiritual challenges.

For contemporary readers, Revelation 12 offers profound theological reflections concerning spiritual conflict, divine protection, human suffering, and eternal hope, resonating deeply with modern spiritual struggles, ethical dilemmas, and eschatological aspirations. Its vivid imagery, symbolic depth, and theological reflections provide invaluable spiritual resources for engaging contemporary questions concerning evil, suffering, divine sovereignty, and human resilience, emphasizing ultimate hope, divine justice, and cosmic redemption.

In conclusion, Revelation 12 stands as a profoundly significant, spiritually evocative, and theologically rich chapter within Christian scripture, offering modern readers powerful symbolic narratives, theological reflections, and spiritual insights. By engaging deeply with Revelation 12, contemporary readers are invited into profound theological reflection, spiritual contemplation, and moral resilience amid humanity's perennial struggles, spiritual conflicts, and eternal aspirations, underscoring the enduring power, relevance, and significance of this extraordinary biblical text.

Revelation 12

The Woman and the Dragon.

1. A great sign appeared in the sky, a woman clothed with the sun, with the moon under her feet, and on her head a crown of twelve stars.a

2. She was with child and wailed aloud in pain as she labored to give birth.

3. Then another sign appeared in the sky; it was a huge red dragon, with seven heads and ten horns, and on its heads were seven diadems.b

4. Its tail swept away a third of the stars in the sky and hurled them down to the earth. Then the dragon stood before the woman about to give birth, to devour her child when she gave birth.c

5. She gave birth to a son, a male child, destined to rule all the nations with an iron rod. Her child was caught up to God and his throne.d

6. The woman herself fled into the desert where she had a place

prepared by God, that there she might be taken care of for twelve hundred and sixty days.

7. Then war broke out in heaven; Michael and his angels battled against the dragon. The dragon and its angels fought back,

8. but they did not prevail and there was no longer any place for them in heaven.

9. The huge dragon, the ancient serpent, who is called the Devil and Satan, who deceived the whole world, was thrown down to earth, and its angels were thrown down with it.e

10. Then I heard a loud voice in heaven say:

"Now have salvation and power come,

and the kingdom of our God

and the authority of his Anointed.

For the accuser of our brothers is cast out,

who accuses them before our God day and night.

11. hey conquered him by the blood of the Lamb

and by the word of their testimony;

love for life did not deter them from death.

12. herefore, rejoice, you heavens,

and you who dwell in them.

But woe to you, earth and sea,

for the Devil has come down to you in great fury,

for he knows he has but a short time."

13. When the dragon saw that it had been thrown down to the earth, it pursued the woman who had given birth to the male

child.f

14. But the woman was given the two wings of the great eagle, so that she could fly to her place in the desert, where, far from the serpent, she was taken care of for a year, two years, and a half-year.g

15. The serpent, however, spewed a torrent of water out of his mouth after the woman to sweep her away with the current.

16. But the earth helped the woman and opened its mouth and swallowed the flood that the dragon spewed out of its mouth.

17. Then the dragon became angry with the woman and went off to wage war against the rest of her offspring, those who keep God's commandments and bear witness to Jesus.h

18. It took its position on the sand of the sea.

Thank You for Reading

Dear Reader,

We hope this timeless classic has sparked your imagination and enriched your literary journey. Now that you've turned the final page, we want to share a vision for the future of reading—one where every classic you've ever wanted to explore is at your fingertips, in a format that best suits your life.

We'd like to invite you to gain immediate, unlimited digital & audiobook access to hundreds of the most treasured literary classics ever written—along with the option to secure deluxe paperback, hardcover & box set editions at printing cost. Together, we can spark a new global literary renaissance alongside our small, independent publishing house called "The Library of Alexandria."

Thousands of years ago, the Library of Alexandria stood as a beacon of knowledge—until it was lost to history. We aim to reignite that spirit of preservation and discovery right now, in the modern age—only this time, it's accessible to all, in every language and every format.

Picture a world where every timeless classic, novel, poem, or philosophical treatise is not only available to read but also updated for today's readers—modernized, translated into any language or dialect, and ready to enjoy in any format you choose, whether that is in an eBook, audiobook, paperback, or deluxe hardcover & box set version a printing cost.

By joining our movement to rebuild the modern Library of Alexandria, you become part of an unprecedented mission to offer:

- **Unlimited Audiobook & eBook Access to the Greatest Classics of All Time**

  Instantly explore thousands of legendary works, from Plato and Shakespeare to Jane Austen and Leo Tolstoy. All are instantly ready to read or listen to, giving you a complete literary universe at your fingertips.

- **Paperback & Deluxe Editions at Printing Costs:**

  Purchase any title in a paperback, deluxe hardbound, or deluxe boxset edition at printing costs, shipped right to your doorstep. Curate your personal library of Alexandria with editions worthy of display—crafted to last, designed to captivate, and delivered straight to your door.

- **Modern translations for Contemporary Readers in all languages and dialects**

  Discover a vast selection of classics reimagined in clear, current language—no more struggling with outdated phrases or obscure references. Next to the original versions, we aim to offer translations in as many languages and dialects as possible.

  As we continue our translation efforts and add new languages, readers everywhere can connect with these works as if they were written today. By bridging linguistic divides, you're contributing to ensuring that these timeless stories become more meaningful, accessible, and inspiring for people across the globe.

- **Your Personal Library of Alexandria:**

  Over the months and years, you'll curate a unique physical archive of classics—each volume a testament to your taste, curiosity, and love of knowledge. It's not just about owning books—it's about curating a cultural legacy you'll cherish and pass down for generations to come.

- **Join a Global Literary Renaissance:**

Your support fuels an ongoing mission: allowing us to reinvest in offering deluxe print editions (including special boxsets) at their true cost, broaden the range of available formats and translations, and extend the reach of these works to new audiences worldwide. By joining today, you're not just preserving a legacy of masterpieces; you set in motion a powerful wave of literary accessibility.

We are more than a publisher—we're a movement, and we can't do it alone. Your support lets us scale our mission, preserving and reimagining history's greatest works for tomorrow's readers.

Become a Torchbearer of knowledge.

Thank you for picking up this book and allowing us into your literary journey. As you turn the pages, know that you're part of something larger: a global effort to keep these stories alive, share their wisdom across borders and generations, and spark a true cultural revival for the modern era.

If this resonates with you—please consider taking the next step by visiting:

www.libraryofalexandria.com

With gratitude and a shared love of knowledge,

The Modern Library of Alexandria Team

Visit:

www.libraryofalexandria.com

Or scan the code below:

9 781804 217962